TOM

MIRACLE
OF
HOMINY FALLS

A TRUE STORY OF
SURVIVAL DEEP INSIDE A FLOODED
WEST VIRGINIA COAL MINE

MIRACLE OF HOMINY FALLS

by Tom Surbaugh
© *Copyright 2017 Tom Surbaugh*
All Rights Reserved.

Sue Adams – Editor
Susan Jones – Graphic/photo layout design

Photo Courtesies:

Shelly Walkup Adkins: Eli Walkup (Walkup family collection)
Sandy Sevy: Frank Burdette (Burdette family collection)
Vicki Rose: John Moore Jr. (Moore family collection)
Mike Davis: Frank Davis (Davis family collection)
Saxsewell #8 Mine map layout: West Virginia Archives.
Telegram: West Virginia Archives

Special Thanks:

To Vicki Rose for sharing her story and her father's hand written journal.
To Robert Deason for guidance.
To Becky Eastling for book layout.

Cover Art by JeshArt

Table of Contents

Prologue 4
Introduction.. 8
Chapter 1 Miner's Morning............................ 9
Chapter 2 Mantrip 20
Chapter 3 We've Hit Water 30
Chapter 4 Head Count 39
Chapter 5 The First Night 57
Chapter 6 Dark Days 65
Chapter 7 Week from Hell 70
Chapter 8 Daylight 75
Chapter 9 Footprints in the Mud 84
John Moore Jr.'s Handwritten Journal............. 104
Notes 109
Index 128

Prologue

Coal deposits occur in fifty-three of West Virginia's fifty-five counties, and twenty eight of those produce coal, though I did not grow up around coal mining. My hometown in the Mid-Ohio Valley river city of Parkersburg sits on the eastern banks of the Ohio River and outside the mountain state's lucrative coal seams. Aside from the occasional barge pushing coal along, most activity on the Ohio River came from industrial plants near home producing chemicals, fuels, metals, and plastics. However, as a child, I was well aware that my grandparents, who lived three hours Southeast, were in the heart of coal country.

My maternal grandfather, Ralph Mullins, lived in Hilton Village a small, unincorporated community near the rim of the New River Gorge in Fayette County. He was a railroad man who had spent his work life hauling coal and timber through the mountains along cliff-side tracks of the Appalachian Highlands.

I first learned of the Hominy Falls tragedy as a teenager when my grandfather asked me if I had ever heard the story about his nephew, Frank Burdette, who had died when the coal mine he was working flooded. Grandpa explained that in 1968, mine officials called him to come and identify his nephew's body.

My paternal grandparents lived ten miles from Hilton Village along the Midland Trail in Rainelle, a small town in neighboring Greenbrier County. My step-grandfather, Clarence Huffman, was a career coal miner. His sister Hilda was married to a fellow miner, Eli Walkup. Eli was working the low coal at Hominy Falls alongside Frank Burdette on the morning of the accident. Walkup, Burdette, Claude Roy Dodd, Jr., and Renick McClung were all killed.

The weekend prior to the accident our family had travelled from Parkersburg to visit my grandparents and the Burdette family in Rainelle. Just only three at the time, I was told that Frank entertained us after dinner by playing the mandolin, an instrument that he learned to play by ear.

Shortly after returning home to Parkersburg, Grandpa Mullins called and informed our family that something had happened at Frank and Eli's mine; they were not sure of any of the details yet, but it was pretty bad.

For our family, and the others that lost loved ones that day in 1968, the miracle of Hominy Falls is a constant reminder of God's grace for the 21 men that survived, and God's promise to the four that were taken home who thereafter rest in His miracle of eternal life through His son Jesus Christ.

MIRACLE OF HOMINY FALLS

Tom Surbaugh

Introduction

Twenty five West Virginia coal miners entered the Saxsewell No. 8 Mine at Hominy Falls in Nicholas County on May 6th, 1968. Fifteen of the miners were rescued five days later after the mine was flooded when a continuous mining machine holed through to an unmapped, abandoned mine filled with millions of gallons of water. Ten other miners were working in a section nearly two miles into the mine. They were considered lost. Rescue workers finally reached the men. Four had been killed. The other six men had survived in the darkness and cold for ten days. This is their story.

Miner's Morning

A spectrum of vivid gold, red, and violet wild flowers beamed from beneath the dew-covered tree line that made up the wooded border of the Moore family property line. The pleasant view from the family's opened kitchen window was filled with budding aroma, a gentle breeze, and bird-chattering sounds. Toward the east the sun was trying to peek over the mountain, and John Moore Jr.'s neatly prepped hillside garden anticipated the warmth of its' rays. It was another picture perfect early May morning in Buckeye with postcard weather blessing the townspeople who inhabited this picturesque region in southeastern West Virginia.

John's twenty-year-old daughter Vicki (now Vicki Rose) had finished breakfast and said her goodbyes to one-year-old daughter Stefanie and her mother Gladys as she headed out the door to begin a normal Monday morning start to the work week. She was a hairdresser at Mary's Beauty salon in nearby Hillsboro. Her trip to work was about six miles from her parents' home, a small and unincorporated community in rural Pocahontas County.

Vicki's mom, who rarely drove following a car accident a few years earlier, stayed at home to watch the baby while Vicki and her dad, a long-time coal miner, went off to work each day.

As Vicki made her way in for an 8 a.m. start at the shop, her father had already arrived at his job site in Hominy Falls in neighboring Nicholas County. John, a shift foreman, would rise every morning at 4 a.m., get dressed, have breakfast, prepare his lunch bucket for work, and be out the door by 5:30. He was usually on his way before anyone else in the house was awake. Vicki often told her father the evening before at dinner to have a good day since he would usually be the first to go to bed. As always, her mother would have dinner ready for him when he got home. It was a repetitive and redundant schedule, but comfortable and safe. "He just had a normal job to us. We never really discussed the dangers or the possibilities of something bad happening, Vicki Rose said. "He had been a coal miner since I was a baby, and that was just the way it was back then."[1]

John's commute to work was over 50 miles one way and he had been working at this particular mine for over a year. He sometimes provided a carpool for fellow miners who paid him 50 cents a day to drive them to work in his Chevrolet Carry-All.[2] But mostly he made the curvy trip through the mountains alone. He left Buckeye on Route 219 south to Mill Point and continued on highway 39 West through Richwood where he connected to West Virginia State Highway 20 after a left turn across a bridge. Another left on highway 20 took him through the town of Leivasy to Hominy Falls, a small hamlet, named for the beautiful falls that majestically flow over the ancient sandstone boulders that mark a dramatic eight-foot drop in elevation along Hominy Creek. It is located in virtually the very center of West Virginia, six miles north of the town of Quinwood and just a few miles outside the western boundary of the remote and rugged Monongahela National Forest.

Of course the drive time varied with changes in the weather. John allowed extra time, especially during winter, to make it on time. Without any complications, the trip usually was about an hour.

It had only been a few weeks since the morning routine included time to shovel snow and scrape windshields. The 26 men that made up the 7a.m. to 3p.m. day shift left home well before sunrise for the still frigid, but scenic Monday commute to their Hominy Falls worksite.

A warmer, early May breeze had cleansed the mountain air and streams ran briskly through the now thawed rivers and creeks in America's birthplace of rivers – the eastern region of the Allegheny Highlands.

West Virginians had endured a winter where record-low February temperatures often dipped below zero. 1968 was a leap year and the coldest month of the year held on with an icy 29th capping February's daily average low temperature at 14 degrees.[3] Appalachia had finally broken from winter's frozen grip and Moore's early May morning drive through the Greenbrier Valley to work was now much more pleasant.

Moore dutifully put in eight hours a day in the mine and had done so for many years. He greatly enjoyed his time at home with his family and had a passion for working in his garden and taking care of his property. He took special joy in singing at church and was smitten by the grins he constantly tried to produce and to receive from his granddaughter.

John Moore Jr. circa 1959

All 26 men on shift arrived for work on Big Sewell Mountain. All but one worked underground. They parked and prepared for the mantrip outside the Saxsewell #8 Mine owned by the Gauley Coal and Coke Company. Fifteen of the twenty six men in one group and ten more working deeper in the mine, including Moore. There was one supply man who operated battery-powered buggies used for deliveries and miner transport. This shift was the first of the week after the mine had sat idle over the weekend.[4] Gauley Coal and Coke was one of many mining companies operating at and near Big Sewell Mountain at that time.

Some miners on shift were local, but like Moore, many had as much as an hour drive to the mine. The men arrived for work from nearby communities like Leivasy, Nettie, and Snow Hill. Others commuted from outlying areas to the mine from small, blue-collar towns like Rainelle and Rupert from the east and Summersville to the west. Their conversations were varied as they gathered together. The ride into the mine was bumpy, noisy, and not well suited for conversation. In the few minutes the men had to catch up from the weekend before boarding, they mostly talked about their kids, working their gardens, and upcoming summer vacations. They rarely talked about work. There would be plenty of time for that when they reached the coal.

Work was steady and consistent. The West Virginia economy was unusually strong and the unemployment rate was headed toward its lowest point in 15 years.[5] Even though mining coal was dangerous, back breaking work, it was there and miners were happy to have it.

Far from the high rise mega structures and chaos of big city life, people who lived in near proximity from the shadows of the ancient New River Gorge religiously went to work. A rugged stretch of topography, this section of the Sewell coal seam has helped provide the United States ample resources in building its industrial strength. Coal, timber, and river power are abundant there.

Heavily travelled local roads like the West Virginia Route 60 Midland Trail remained off the beaten path by comparison to America's suburban landscape littered with interstate and super highways. Many workers at Hominy Falls and nearby mines like it, commuted every day using roads like Route 60, a windy, mountain-rim road known to be the oldest in the United States that runs along the banks of the New, then Gauley Rivers and ends at the industrial edge of east Charleston at the start of the Kanawha River. Workers turned on seldom used exits to state highways and some single lane tracks to the work sites of some of the Nation's most remote and unforgiving coal mines.

In a year that would go down as one of the most turbulent in United States history, the first quarter of 1968 in West Virginia went virtually unnoticed. The exception, daily local newspaper scans for familiar names and faces of those just a year or so removed from area high schools that had been killed or were missing in action in the Vietnam War. Per capita, West Virginia had the highest casualty rate of its citizens in the South East Asia conflict.[6] 84.1 percent for every 100,000 men, more than any other state.[7] The national average was 58.9 percent. Over 36,000 people from the Mountain state served in Vietnam during the war.[8] Most West Virginians knew someone or were aware of someone from their area who was serving in the United States armed forces during the conflict. Among them, Vicki Moore's Marlinton High School classmate John Ray "Chipper" Williams, an Army Sergeant, who lost his life November 29th, 1968, in the jungles far away from home.[9]

The late January 1968 North Vietnamese TET offensive had the United States military fighting in the streets of Saigon. The U.S. embassy was invaded and held for six hours. The enemy had taken the fight from the jungles into the city, marking a major turning point for America's attitude toward the war.[10]

The United States involvement in Vietnam created great tension on college campuses around the country. Many young West Virginians were protesting the war. Students in large numbers, including those at the Mountain State's largest institutions, West Virginia University in Morgantown and Marshall University in Huntington cursed the establishment.

With tensions already at the breaking point over Vietnam, civil rights leader Dr. Martin Luther King Jr. was assassinated on April 4th in Memphis. Ensuing riots in some of America's largest cities had the nation's population centers reeling amidst civil, social and political unrest.

A point of unification and a common goal shared among many Americans during that era came from the National Aeronautics and Space Administration (NASA) and President John F. Kennedy. His early

1960s decree challenged the space program to put an American on the moon by the end of the decade.[11]

The 1968 Apollo 7 and Apollo 8 missions putting the first manned space crafts into orbit took the space program one step closer. The country and world's attention tuned-in for each event as NASA continued to clear one hurdle after another in completing the late president's goal.

Toiling in three-foot tunnels thousands of miles from war, hundreds of miles from social and civil protest, and worlds away from the outer limits of space, mountaineers steadily mined coal. Their tireless work stirred the country's continued industrial growth and its military and space exploration efforts. At Hominy Falls, harvesting coal from Big Sewell Mountain was business as usual.

The men on Saxsewell No. 8 Mine day shift prepared for the first ride of the week into the darkness. They climbed into the mantrip buggies wearing steel-toed boots and knee pads. Battery-operated headlamps provided light leading their way through tunnels to the coalface. This was their routine working a "deep" mine. This was life working the low coal.

For John Moore it was like any other Monday to begin the work week. But this May 6th would turn into a day that would change his life and the lives of his crew and their families forever.

Mantrip

Coal generates roughly half of the electric power produced in a year in the United Sates. In West Virginia, coal provides 99% of generated electric power. Coal is the majority source of electric power in 32 states.[1] Like their mining fore-fathers, West Virginia miners were eager to play their part in keeping America and specifically their state's electricity flowing.

Deep in the heart of the mountain state's south central Greenbrier coal seam, the Gauley Coal & Coke Company began mining operations at the Saxsewell Number 8 Mine at Hominy Falls on November 7, 1966. The mine produced an average of 1,200 tons of coal per day. By May, 1968, 68 men, 66 underground and two on the surface were employed on one maintenance and two coal producing shifts a day, five days a week.[2] The coal was highway transported by Auto Trucks and on rail by the Baltimore and Ohio Railroad. Trucks from 10,15 up to 20 ton capacity moved the coal locally, usually within 10 miles, to bigger tipples where it would be poured at railroad platforms into coal cars for shipment.

Employees came locally and from surrounding counties and were rewarded with solid wages and additional benefits.

A shift started with the mantrip, a term to describe the miner transport into the mine. It is often long and dangerous in its own right. Just as the sun came up, the day shift crew entered the mine right at 7 a.m. after boarding vehicles sometimes called Keysers. These machines consisted of rubber-tired mine cars hauled by battery-powered mine tractors.[3] Depending on the work area, the mantrip could take the first 30 to 45 minutes and sometimes up to an hour of a miner's shift for a mine that had been in service for any length of time.

Ernest Fitzwater, a supply man and transport driver, delivered the men to designated mining locations without incident and all conditions seemed normal. The first group of the two continuous mining machine crews consisted of 15 men performing normal operations in normal conditions 3,500 feet from the mine entrance portal.[4] The other group of ten miners continued to two right turns off the south main tunnel, approaching 4,700 feet in distance from where they boarded the transport vehicles.[5]

There are a variety of different kinds of coal mines and West Virginia mines vary. The three basic types of underground coal mines are shaft, slope, and drift mines. The type of mine used depends on the depth and location of the coal seam and the surrounding terrain. Shaft mines are generally the deepest and have vertical access to the seam using elevators to carry miners and equipment to work areas. Slope mines, which usually are not very deep, are inclined from the surface to the coal seam. The Saxsewell No. 8 at Hominy Falls was a drift mine. Another name for this type is deep mine. No. 8 mine had a horizontal entry into the coal seam from a hillside. This was high-grade, bituminous low coal. This prehistoric vegetation had been compressed by soil, water and the elements for over 100 million years. It was valuable fossil fuel several hundred feet below the surface and only three feet wide up from the mine floor. The rooms the men were working in areas deepest into the mine carried the temporary names Number 1, 2, and 3 of the continuous miner machines cutting them. They arrived to these locations at 7:30 a.m.[6]

The men spent most of their eight hours crawling on their hands and knees in man-made tunnels. Their heavy work gloves, knee pads, and steel-toed boots took a beating. They were navigating in the deepest part of the mine and all of the dense quarters well beneath the surface were cold and damp. These low mines in the southeastern region of West Virginia are considered much more dangerous than the mines of the southern region of the state towards the mines fields that stretch cross the border into Kentucky.

The southern mines have an average tunnel height of six to nine feet. Miners spend more time standing up, and mobility is much greater. In low coal, miners on hands and knees have much less time to escape fire, poisonous gas accumulation or flooding.

Moore was the "face boss" and in charge of the men of his crew on this shift. He was 46 years old and like many of his fellow miners, he had never worked in a mine where he could stand up. The years of back-breaking work had taken its toll on just that, his back. His spine was riddled with arthritis and the pain was sometimes unbearable. It had become increasingly difficult to work in such cramped, closed-in surroundings year after year. He was working his 19th year and closing in on the 20-year retirement mark. Relief was in sight. No more bending over, navigating the sub-terrain in dim light. No more breathing coal dust or dealing with the dark and the cold. He made few complaints. Pain and discomfort were part of the job.[7] Although dangerous and physically exhausting, these were good jobs in a state where high paying jobs were rare. The federal minimum wage was increased in February 1968 to $1.60 an hour.[8] Most miners in West Virginia made about $27 a shift depending on their experience and what position they held, which was more than twice the minimum wage.

Well before the morning work began, Charles Beam, foreman on the previous, and last shift Friday night, had already made a visual inspection of the mine. Always damp, one section had more water than usual on the floor. This section was in new production and dog-legged to the left beyond the farthest reaches of No. 8 Mine. Beam thought little of it as sitting water was common after going undisturbed over the weekend.[9]

Several of the employees that worked the face, coal in the direct path of the continuous cutting miner, stated that water had seeped through the coal near the face of Number 1 Room on the last work day, Friday, May 3rd and some water accumulated near where the men were preparing to begin mining. None of the employees or officials was particularly concerned as they were all of the opinion that they were in solid coal.[10] By the time the men reached the work area, Moore witnessed a small stream of water flowing from one of the newer section's three rooms. The Number 3 Room was slowly taking on water, but that was not unusual. Work commenced and that crew began mining the day's coal in the Number 3 Room.[11]

Although moisture and flooding inside a coal mine are very dangerous and monitored… The most obvious concerns for potential disaster are explosions from ignited coal dust, methane, and flammable gases or roof cave-in. Records in West Virginia up until 1964 for coalmine on-the-job fatalities only considered it a disaster if it involved deaths of five people or more. Of the 101 disasters recorded from 1884 through 1966, all but nine were explosions, fires, or lethal exposure to gas and poisonous fumes causing suffocation. The other disasters included falling cages, haulage accidents, run-away mantrip cars, and roof falls.[12]

Even though West Virginia has a long history of deaths in coal mining up to that point, there had been no recorded disasters in the Mountain state with deaths due to mine flooding.[13] The state holds the unfortunate claim to the worse mining disaster of any kind in United States history when multiple explosions in 1907 at the Fairmont Coal Company's Monongah Number 6 and Number 8 mines killed 361 people. That number is believed to be an estimate and the number of deaths was more likely close to 500 with many more men (and boys) working in the mine than was officially documented.[14] (Employment records were not often kept up to date in 1907 and many miners who were never found or accounted for had been working undocumented and not on any official company books).

Almost halfway through 1968 things had remained quiet, at least for a while. There had not been any fatalities in West Virginia mines in over 18 months, a stretch that included all of 1967. The most recent fatal incident in West Virginia mines to that point had happened on September 10, 1966, at the northern panhandle Tridelphia mine in Ohio County when four men were killed in a mantrip accident at the Valley Camp Coal Number 3 Mine. The most recent accident that claimed lives in a mine in the region near Hominy Falls happened on July 23, 1966, at Mount Hope in neighboring Fayette County. Seven miners died there in an explosion at The New River Coal Company's Siltix mine.[15]

Moore joined the others in Number 1 Room where they had been repairing their own mining machine, No. 5, which had a "paddle chain" that was not functioning properly and needed repair.[16] The continuous miner in Number 2 Room was not in operation that morning and no one was working in that section at that time.[17]

Adjacent to the Saxsewell work site was one of the abandoned mines known to the workers as the McKenzie mine. This mine was known as an "old timer" first operated in the early 1900's. There had not been any operation in that area for over two years.[18] The maps and mining guides for the No. 8 Mine did not include the old McKenzie Mine workings. Without a proper and updated survey of the area, the convergence of the active and inactive mines was now being separated by a dangerously small amount of coal serving as a very narrow barrier.[19]

This was an example of the ever-present judgment battle for coal companies walking the fine-line between profits and safety. As long as the conveyor belts were moving coal and railroad cars and transport trucks were being filled and making shipments, cutting corners where miner safety is concerned was more often the rule and not the exception. Work on a shift was rarely held up due to a lack of mine maps or the latest updates from engineers. For coal companies, time is money. When men are not at the face and machines are not cutting and loading coal, money is lost. And coal brings big money.

Mine maps had not been updated and a recent copy of map inspections and updates was not available in the mine foreman office.[20] The abandoned old timer had become a holding tank for millions of gallons of acidic water being held back under extreme pressure.

At 9:40 a.m. back in Room 3, the continuous mining machine, a Jeffery L-100 boring type, was about to complete its third sweep of the morning.[21] Eli Walkup was at the controls of this relentless power continuously tearing away coal at a feverish pace across the face. Unknowingly, the miners, especially Moore and the other nine miners on shift working the deeper sections, now had just a thin sheet of coal separating them from the source of all that water.

We've Hit Water

The Sewell coal bed running through the Hominy Falls property was leased by the Gauley Coal and Coke Company from Eugene and Thomas McKenzie of East Rainelle, West Virginia, in April 1962. The McKenzies had operated the No. 4 Mine, Sugar Grove Coal Company, prior to their leasing the tract of coal, and they retained part of the property for five years to continue operating No. 4 Mine. The McKenzies operated the No.4 Mine until 1963, and several different contractors mined coal on the reserved boundary until March 1966, when the No.4 Mine was permanently abandoned. After acquiring the property, the Gauley Coal and Coke Company would lease tracts of coal to small operators in addition to operating the Saxsewell No. 8 Mine.[1] Many of the miners working the two shifts at Hominy Falls had worked at many different mines throughout the area over the course of their careers. There were nine active coal mines at Big Sewell Mountain in 1968.

Number 8 Mine held the medium-volatile Sewell coal-bed that averaged 32 inches in thickness locally.[2] All coal was loaded mechanically.[3] Mine maps prepared for Number 8 were posted on the wall in the mine foreman office. Usually, the engineering staff of the Saxsewell mine extended the map posting, in the mine office upon completion of their weekly surveys.[4]

The coal industry of West Virginia is over 150 year's old. It is estimated that over those years, thousands of mines were abandoned. In many cases, very little information was recorded on these mines. No laws were on the books requiring maps to be maintained after a mine shutdown. As early as 1883, mine maps were required to be furnished to the inspector. But when a mine was abandoned, it was basically forgotten. It was not considered that future problems could arise if no record remained.[5]

Franklin Davis, Gauley Coal & Coke Company mine superintendent, stated that the posted mine map was usually three to four days behind. The engineers had not visited the Saxsewell No.8 Mine for about ten days. They had not been brought up-to- date in 1968. Mapping did not show or indicate abandoned mines in close proximity to the active Saxsewell mines' two right entries off the main (tunnel).[6]

Five of the men were working that morning in section Number 3 Room. It was located on the left past the second entry passage right off south main entries.[7] Coal miner directions for the road map that gave precise location for cutting specifications and preparation for conveyor belt loading and removal. Coal from the continuous miners was discharged onto conveyors that relayed it to a chain conveyor on the entry then in turn discharged the coal onto a rubber conveyor belt a short distance from the cutting zone.

The other five workmen in Number 1 Room had about the distance of eight pans (pan/6 feet 2 inches) to slab off the pillar before completing work in the space and going on to Number 2 Room to resume mining there. The work plan that morning was typical of most days in the mine. They would cut into an area leaving 75 feet in center of solid coal, drive them up 40 pans, cutting 30 feet of the coal away in advance. The crew then would turn the miner around and 30 feet of coal would be cut out of the pillar as the massive machine returned to the entry, leaving about 15 feet of solid coal remaining in pillar.[8] This in part was to keep the roof from collapsing while mining as much coal as possible.

At approximately 9:40 a.m., timberman Gene Martin was standing on the right side of the Number 3 Room near the tail of the bridge conveyor as the continuous miner was completing its third pass across the seam cutting left to right. The cutter had advanced the place about 110 feet and reached about 12 feet in depth when Martin heard fellow timberman Frank Burdette working on the right side of the mining machine yell, "Boys we hit water!" Martin turned toward the coal face, toward Burdette's voice when a wall of water slammed into him.[9]

It's pitch-dark. Water is violently rushing in. The three- foot high tunnels filled fast, and the water was rapidly getting deeper and deeper. The sub-terrain deluge swallowed the men. The coal-cutting equipment, steel chains, and the heavy conveyor pans and men were all tossed together in the churning tidal wave. The awesome force of in rushing water moved the massive 12 ton continuous miner about 17 feet away from the face and washed pans and the chain for the chain conveyor line out of the room and piled the very heavy equipment in a heap in the Number 1 Room entry. Martin was knocked back and out of Number 3 and toward the mouth of the Number 2 Room.[10]

The vast darkness that exists in such cramped and tight spaces within a mine are normally enough to bring caution and even fear to the bravest and most experienced that work in mines. Add tidal quantities of water rushing in and no place to go, and it becomes a torrential horrifying nightmare.

The men were going through all this and then some. Experience and training bring a sense of control in managing the constant dangers of working in a coal mine. The biggest challenge to overcome is the fear of not letting what is not known take control. They may be able to handle the natural elements of coal mining. Cold, dark, dust, dampness, and the threat of roof collapse are always present. Physically exhausting labor in poor quality air quickly takes its toll. Many who try to take on the coal become 'former' miners. Coping with the tricks the mind tends to play on someone working shift after shift in a moving office that sits thousands of feet below the earth's surface. It's inevitable that this very stressful environment will force many to find another line of work. For the career miner, those that display the mettle to conquer all the dangers and short falls of such a vocation, a co-worker brotherhood develops. One of the perks in an occupation that carries the realization that each work day there is the possibility one may not make it back home. Poor visibility, bad air, trapped gases, roof collapse and flooding make up many of the ingredients for danger that are present in this treacherous occupation.

The bond between fellow miners is similar to the fraternity of military soldiers who experience combat together. People who have never ventured toward the earth's core after coal could ever understand this camaraderie. Vicki Rose said her father took pride in his responsibility as the foreman. "He looked after the men the best he could. He carried a small, pocket-sized notebook to keep track of not only supplies he would later order, but the miners' hours, time off requests, and other personal reminders and notes from the men." Moore had worked his way through the ranks to become a foreman. He was very hands on and wouldn't ask anyone on his watch to do something he wasn't willing to do. "He never handed out orders and then stayed on surface." His daughter said, "He went down to the low coal and worked side-by-side right along with everyone under his charge. He always seemed to be well liked by the men, especially those that rode with him to work from time to time."[11]

Frank Burdette (1967) a few months prior to the accident.

But no training or on the job experience could have prepared Moore or the miners with him for what had just happened. The other four men working the Number 3 Room with Martin were carried completely out of the working space by the water. They had no chance. Burdette, miner helper Claude Roy Dodd, Jr., timberman Renick McClung, and continuous miner operator Eli Walkup were all drowned.[12]

Eli Walkup circa 1967.

Eli Walkup was one of many coal miners that served in the United States military.

The inundation left the five men in Number 1 Room desperately searching for a way out through the tunnels that worked up and down. Foreman Moore, along with miner operator Jennings Lilly, miner helper Edward Scarbro, beltman Larry Lynch and electrician Joe Fitzwater, were working the Number 5 miner in Number 1 Room when they heard

37

the shouts of the Number 3 Room crew the moment that crews' miner holed through. Moore's crew went directly to the mouth of the Number 2 Room and got a first look at the massive amount of water that was pouring through with great force. It had completely filled Number 3 Room.[13]

Moore, Lilly, Scarbro, Lynch, and Fitzwater found Gene Martin lying near the entrance of Number 2 Room and helped him from the water. A sharp object had badly injured Martin's hand during his submerged tumble when the water's impact engulfed him.[14] He had momentary been knocked out. But he was alive. The men collected Martin and tried to make their way through the water but were unable to find a clear path back to the main coal conveyor belt and a way out.[15] The three-foot high tunnels were no match for so much water.

In just seconds, the acidic water had risen to an inescapable level.[16] The men had no choice but to retreat back to Number 2 Room that sat at just enough elevation - a foot or so above the water line - where they could find safe footing away from the underground river.[17]

Head Count

With no way to move forward the men backed out and slightly up returning to the Number 2 Room at the highest possible elevation. They instinctively gathered any belongings they could and collected some brattice cloth (a fire-resistant fabric, used with brattice partitions in mining as a way to help direct air flow for ventilation).[1]

The rushing water was still on the move and rising within an inch from overtaking their location that was just out of harm's way. The men began to pray. Larry Lynch, a part-time minister, fell to his knees and prayed. Moments later the water stopped rising.[2] The men immediately began to build a make shift shelter by constructing a five-feet wide and roughly eight feet long structure near the face of Number 2 Room. They used three-foot layers of the brattice cloth and some timbers as barrier material to gain some separation from the water.[3] Their refuge on the No. 2 room ledge was just a bit larger than the dimensions of a bed on a pick-up truck.

Through the confusion and scramble to find shelter and get out of the water, Moore's loose-leaf notebook stayed securely in his jacket pocket. As the foreman, he responsively began to write down what was happening:

"*May 6, miner cut through to old mine which was full of water. We tried to get out to the belt but couldn't make it. The only chance we had was to go back to the two rooms. We prayed. L. Lynch fell on his knees and prayed and the water stop[ped] rising. We put up brattice and set timbers and made us a place to stay in. Scarbro and I got 3 dinner buckets and 3 coats. But one bucket was full of water and we couldn't eat anything in it. We had two buckets which had about 4 sandwiches in them, 3 candy bars, ½ gal. of water and 1 pt. milk.*[4]

TWO dinner buckets⋯SIX men. The life-blood of a coal miner is contained within the aluminum confines of a shoebox- sized dinner bucket. Other than the freshly laundered clothing they wear at the beginning of a shift, a durable, clean dinner bucket is their link to life out of the mine. It is a miner's connection to family, home, and why they were doing what very few would do for a living. They had to put food on the table and in turn that same food was carried each day into the mines, continuing the cycle.

Coal mining is work hard. In the early days, miners crawled on their knees and used heavy steel picks and shovels. They were paid by what they individually mined. The coal was loaded manually onto mule- drawn carts or even carried out on foot in sacks. Miners were paid wages based on weighed tallies. Despite such improvements and innovations of continuous mining machines, mechanical loading and removal, and regulations requiring more air ventilation, coal mining in the late 1960's remained difficult and back-breaking work.

Miners greatly look forward to their mid-shift break. A well-stocked dinner bucket and thermos bottle were constructed with care and packed to sustain any fatigued miner's appetite until the conclusion of a shift, including the mantrip ride out and commute back home. This was what the six men trapped thousands of feet inside the mountain had to ration. Food from two buckets (jelly sandwiches and candy bars), meant for two men, enough for one work day.

The men hunkered down and other than the

strange sounds deep in a coal mine of water splashing and spraying just a few feet away, the small space they now occupied was completely silent. The men suddenly felt the weight of the situation and as a group they all had an uneasy feeling that they could be there for a while. They finished setting some timbers and the brattice cloth. They started to pray.[5]

Meanwhile, the other 15 men working closer to the entrance in the mine did not realize anything had happened for about ninety minutes.[6] The men making up this group were all from West Virginia. Mine superintendent Franklin Davis, 43, of Richwood led the men. Working alongside Davis were brothers Ottie J. Walton, 39, of Quinwood and Andy H. Walton, 36, who lived nearby in Hominy Falls. Another set of siblings mining that section were Osmand L. (Ottie) Dillon, 44, also from Hominy Falls and his younger brother Oscar Dillon, 35, of Nettie. The rest of the crew was made up of Edward F. Rudd, 39, and Glen Amick, 26, both from nearby Leivasy. Amick and Harry Bess, 26, from Richwood were the youngest members of the crew. Elwood O'Dell, 40, came from Crighton. Eldon J.

Collins and Lonnie C. Bennett, were both 32 and from Fenwick. Rounding out the crew were Hershel E. Seabolt, 38, who lived in Craigsville, Roy L. McClure, 47, of Richwood, Addison A. Copen, 57, from Rupert and the eldest miner on shift Isaac L. Casto, 59, who also lived in neighboring Nettie.[7]

The water had not reached their section yet. Their work continued and they had no idea their escape route out of the mine had already been breached by in rushing water.[8]

Just before noon, Ernie Fitzwater was sitting atop his mine buggy about 800 feet from the entrance. He was the closest to the outside of the 26 men on shift that worked underground. He was about to return with supplies to the work areas where he had dropped off the men at their designated working areas that morning.[9] As he turned the battery-operated vehicle and the two trailers used earlier to carry the men, he headed back toward what he and his brother called a "bad hole."[10] His brother Joe was among the ten men he left at the far stretch of the mine nearly two miles from the entrance. The Fitzwaters made up the third set of brothers working for Gauley Coal and Coke at No. 8 mine that morning.

Ernie's brother had been having "bad feelings" about the mine and the brothers had discussed the presence of water evident on the mine floor of where Joe had been working. "Bad" feelings are common for miners who are always on alert for trouble. But Ernie didn't like the feeling he was getting when he traversed back through the mine earlier toward the surface. He knew his brother better than anyone, and he could see concern on Joe's face that morning and it made for a long ride out. He had the thought that he always tried to suppress that this could be the last time he would see his brother alive. Thoughts such as this being an occupational hazard of working the same shift with a family member in one of the world's most dangerous jobs.

The brothers were used to working bad holes. They had been in low mines for years and they both understood that getting out when trouble came would not be easy, especially when any escape route would require crawling through tunnels with just 32-inch openings. They envied miners who worked in other parts of the state in high mines. "Low mines, "they're just bad business to start with. Yes, he thinks again, but the pay is good. The pay is damn good. Nearly $30 a day. In West Virginia, that's damn good money." Fitzwater said. [11]

Fitzwater cranked up the vehicle as his headlamp cut through the darkness to begin his supply run. He quickly noticed damp spots on the conveyor belt moving in the opposite direction. The belt moved continuously carrying coal to the surface for loading. Sometimes the belt would be completely full for long stretches of a run. Other periods would force the belt to run empty during down time for maintenance, safety checks, or equipment failures. Fitzwater became very adept to watching the conveyor for any signs of anything out of the ordinary. He saw water and lots of it. He then noticed pieces of fallen lumber. Fitzwater remembered why his brother had been uneasy in the "bad hole." The men had been seeing water accumulation on the mine floor. This is somewhat common, but they were noticing too much - and every day. [12]

Fitzwater felt sick as his first reaction was to turn and get out of there. Knowing his brother and the nine men with him would most definitely be trapped and probably already underwater. He started to turn the vehicle around in the tight space, but the trailers jackknifed.[13] He jumped off the buggy and was immediately hit in the back with an on-rush of water.[14] A wall of water surrounded him. It was dark, and he was confused. The water began bouncing and tossing him against the sides of the mine walls. He grabbed and took a hold on the conveyor belt still running outward. But it was underwater. He thought if he rode it out he would drown and let go. The water tumbled him about and he eventually did grab ahold of the belt again. He placed a piece of cloth from his hip pocket over his mouth and held on, trying desperately to keep his head above the water. "Move belt!" he prayed. And the conveyor belt ran up and out until he saw daylight. [15]

Fitzwater jumped off the belt once outside and ran to mine supervisor Frank Davis. Edward "Bozo" Rudd, a miner helper, joined Davis. Out of breath, Fitzwater explains desperately…"Water"…"a flood"…"in the mine"…"the others, they're all trapped."[16]

Davis told Fitzwater to get in touch with the Gauley Coal and Coke's general manager, T. A. Salvati and have him begin the process of getting pumps. Davis and Rudd quickly made their way back into the mine.[17] Fitzwater was drenched and stood stunned in the sunlight. "Joe⋯" "My brother Joe⋯the others⋯what has happened to them?"[18]

With Davis and Rudd on the move, word outside the mine travelled fast. Ottie Junior Walton was working with 12 other men about a mile from the entrance and got a call on the Femco radio phone. A voice on the other end commanded⋯"water in the mine, better get out quick." Walton turned to his crew and told the men to stop working. "Let's get out of here!" Walton shouted. [19]

The group heard a noise, then a voice. Davis and Rudd were headed toward them making their way through the tunnel, crawling with water around their knees and wrists as they eased along the mine floor. Davis ordered Walton and the other men to get to higher ground. "Can we make it out?" Junior Walton asked. "Not a chance. That water is comin' in here like a river," Davis said. Davis told Walton they had seen the water topping out to the ceiling and that they would have to wait it out until the water receded.[20]

"There's 10 others down the side shaft. We're going after them." Davis said. Davis and Rudd left the first 13 miners and moved slowly down the tunnel towards the deep side shaft. Junior Walton looked down and was in about eight inches of water. He moved the men to the far edge of the "swag", a long, deep dip in the mine floor. Although the tunnel was less than three feet high, in the swag it's about six feet and a miner could stand up as long as they were less than six feet tall. [21]

Davis and Rudd crawled toward the passage to see if they could find the source of the water. They reached the turn-off that led to Moore's section when the two men met the rushing underground river. They inched forward on their hands and knees toward the water's source at the second right entry off the main tunnel to see if they could get a better look. Moore and the other nine men that went to the far reaches of the mine that morning were cut off. They made it as far as they could, but the water filled the passage all the way to the ceiling. They could not get anywhere near Moore's group's section. The water was relentless and crashing toward them. Reality sunk in and Davis turned to Rudd, "They didn't have a chance," he said.[22]

Davis and Rudd retreated to try and rejoin the other 13 men, but they were trapped at the second south belt tailpiece in the first room of that entry way. The water was rushing by them at lower elevation in the Number two and three entries. The other 13 men were also trapped. They were at the head of the first right entry way directly at the end of the main tunnel. Fortunately, communication to the surface still had not been cut off for these 15 men. The telephone lines to the outside were working.[23] Davis and Walton stayed in contact with mine officials from the outset of the ordeal and worked the situation with company and state personnel on the surface.

After only a few hours, mine officials and rescue operations had seven large pumps prepared to begin the process of getting the massive amount of unwanted water out of the mine as fast as possible. In planning for pump placement, officials took into account at their best estimates as to where the men were located and tried to plan accordingly.

Looking down on the mine from a bird's eye view, the layout was shaped like a lower case letter "h" with the entrance located at the bottom left hand side of the base of the letter. Superintendent Davis and miner Rudd were trapped for the first three days in a space at the second right belt head about a mile and a half off the entry. They joined the other thirteen men when the water receded. That group was off another right entry at what would be the top of the small letter "h," a mile from the entrance.[24] The ten men who were out of contact were working at the bottom right foot of the small "h" diagram nearly two miles from the mouth of the mine.

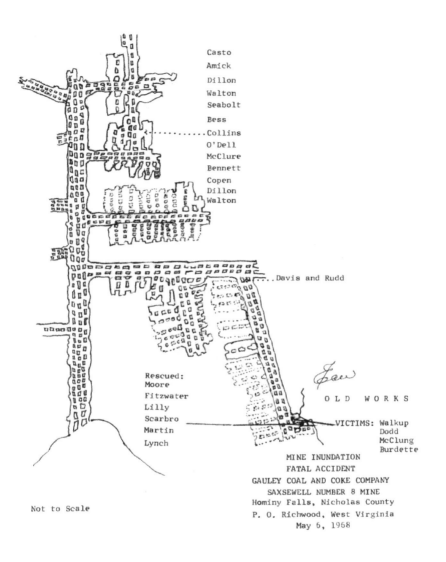

Map of Saxsewell No. 8 mine, Hominy Falls, W.Va. 1968.
Courtesy West Virginia state archives.

In coal mining, any word of trouble travels fast. Nearby mine operations of other companies stopped working. One crew from Imperial Smokeless Coal was trying to rescue the crew of 15 by drilling through their own mine in the same vicinity of the Saxsewell works.[25] Many within the mining community and rescue personnel first on scene already had doubts based on reports of how much water they were dealing with that there were any scenarios the miners from Moore's group could have survived. West Virginia State Mines Chief Elmer C. Workman said some of the miners may be reached "sometime during the night." As for the men deeper in the mine, he said only: "We fear the worst."[26]

No one would admit it but just a few hours from the initial reports of flooding in the mine, the rescue of the ten in the lower region was deemed unlikely and would eventually become a recovery operation.

As officials scrambled to locate and put into service as many pumps as they could make operational that afternoon, John Moore's wife Gladys was having a normal day taking care of her granddaughter. She had just looked in on the baby when off in the next room she heard the television station break into programming with a news bulletin. She heard something about a mine accident in nearby Nicholas County. She listened intently as they reported the confirmation of coal miners trapped due to flooding at the Gauley Coal and Coke Saxsewell mining operation at Hominy Falls.

In Hillsboro, Vicki had returned to the shop from lunch and had just started working on her next customer when her mother called. By then it was about 2 p.m. and it wasn't unusual for her mother to call her at work. Her mom had been babysitting her daughter almost every day since she began working at the salon six months earlier. "Mom and I talked on the phone while I was at work almost every day, but I did not expect a call like that of course. She was not in a panic or anything and actually seemed pretty calm under the circumstances. She had heard on the news that something had happened at dad's mine and we needed to get over there. Mom didn't know many of the details but when flooding and coal mines are mentioned together- we knew it was serious. So I hurried home, picked her up and we drove over to the mine. My grandmother stayed with the baby." [27]

"The crews working at Hominy Falls were relatively small compared to some of the mining companies around the state. With dad being a very hands-on foreman we knew that he would be right in the middle of whatever was going on in there. It took us about an hour to get there and mom didn't say much on the way even though I kept speculating what may have happened. I tried to stay positive. When we arrived, people had already gathered. Some people from a local church had set up a couple awnings and everyone was waiting to find out what was going on." [28]

The different family members of the Hominy Falls' crews did not know each other very well because the stretch a miner would stay with one company at one job site varied. Miners tried as best they could to stay close to home and with a company as long as they could gain the most from their benefits. But miners moved around the region to where the work took them and usually where they could maintain steady employment and make the most money. "My mom talked to some of the other wives and they tried to figure out what was going on. There were a lot of unknowns so it was mostly just waiting and being there to support each other," Vicki Rose said.[29]

"The family area consisted of the awnings and makeshift tents in a level area on the side of a hill not far from the mine opening. Area churches sent people to help, and volunteers brought food. A mine official finally came over and told the families that pumps had begun working and that they wouldn't know anymore until the water levels receded. Everything after that was dictated by the water levels and how far officials and safety personnel could safely go into the mine. That became our routine for the next ten days. We would go over to the mine every morning, stay all day, and drive back in the evening."[30]

Mine Superintendent Frank Davis' son, Mike, found out about the accident on the radio. "I was 18 and that Monday I was driving home from Morgantown at the end of the semester of my freshman year at West Virginia University. I was with my friend Tom Hackney and when we were getting close to home, we picked up WVAR radio out of Richwood and heard about a mine flooding accident at Hominy Falls. I didn't know exactly where my dad was working. When I arrived home, my mother was lying down on the couch very nervous and I knew immediately that dad was in that mine." [31]

It was worse for Mike's sister Dinah. She worked in Morgantown at the WVU president's office. Her husband Joe Rivituso was also a student and when she heard about the accident they left for home. "We all shared an apartment and we didn't have a phone," Mike Davis said. "So Dinah borrowed our neighbors phone to call home. She and Joe left for Richwood and on their drive they heard a report on the radio that there were no survivors. So they thought dad had died." [32]

The First Night

Water pumps were in place and continued working; rescue procedures were spelled out as the first evening drew to a close. Work on boring a delivery hole and system needed to be in place to send food, water and blankets to the 15 men that stayed in telephone communication. Initially, on May 7th, sections of a four-inch plastic pipe were filled with these supplies and sealed with plastic, securely fastened to the main belt conveyor, and sent back to the entrapped men at the main entries. This procedure was used until a nearly six-inch bore-hole was drilled and broke through to the men the following morning at 2 a.m. The materials were then lowered from that point on until the men could be reached.

By 3:30 p.m. the first evening, a three-inch bore-hole reached the coal bed near where Davis and Rudd were stranded. An air compressor was put into operation through this bore-hole for the purpose of ventilation and was operated for 20 minutes out of every half hour. A bigger 5-5/8 inch bore hole reached these two men near the two right south main belt drive shortly thereafter. The men were supplied through this borehole until they rejoined the other 13 men three days later on May 10th.[1]

Moving to join the others was made possible by continuous pumping which lowered the water sufficiently in the south main entries to allow Davis and Rudd to move. Complete rescue of these men would take time, and there were still many unknowns. The severity of the flooding and depths of the water were still not completely known to officials.

For the six men hunkered down in the Number 2 Room, they literally remained in the dark. They were unaware of any rescue operation that had begun, and they had no idea of what had happened to their four co-workers who had been completely washed out of the Number 3 Room. All they knew was that they were damp, cold, and the small space that the six of them occupied was filled with darkness. The men made it through the night and Tuesday morning they began the routine of checking as accurately and safely as they could to measure the water level. Moore wrote in his notebook: *May 7. Water went down about 8 inches. We were all still wet from Monday and pretty cold. We keep checking on water and prayed and tried to sleep.*[2]

With the water level dropping moral increased and the men wondered if it was only a matter of time until they would be rescued. Back ground level at the mine entrance, wives and families of the miners were back from a long, fearful first night. The kind of night that coal miner wives try not to think will ever come. They continued to wait, worry, pray, and hope for any word.

By now, most had heard of the accident on the radio and some had learned about what had happened on the previous evening's local area newscast. Few among the miners' families had been told much of anything the next morning by mine officials. Another funeral home set up a tent that sat empty, giving those who arrived the solemn notion that its intent was to serve as a make shift morgue.

The news circulated that officials were in contact with 15 men by telephone. Members of the media, both local and some national began to come to the mine in droves. The story of trapped miners and organizing a rescue operation quickly gained momentum. Newspapers began publishing photos of the happy wives that now were joking about all of the overtime that their husbands were putting in.

And they teased about their husbands being on a break from them. "Don't you be gettin' into any meanness down there," one joked.[3]

Mike Davis said he was very excited to be able to speak to his father Frank on the phone line that they had dropped from the mountaintop. "It was great getting to hear his voice and I was so happy to tell him I had gotten a job that summer with the National Forest Service. I went with my mother and my grandfather to the mine site every day. After we were able to talk to my dad, we didn't worry about him making it out. It was just a matter of time." Davis said. [4]

By contrast to those thrilled of the news that their loved ones would be okay, the families of those still cut off were left to wonder. Any happiness they felt for those who had made contact quickly was replaced by their own, real fear that they did not know where, how, or when they would get any news on their loved ones.

"We were grateful that they had contacted some of the men, but I just remember thinking that all I wanted was to see my dad." Vicki Rose said.[5]

A few newspapers didn't even refer to the missing 25 men anymore, only 15. The others were now called "the lost ten." But rescuers had a different attitude than the media. They were not about to give up on the others. They now concluded what the men huddled in No. 2 already knew. The men had accidentally cut through into an abandoned, water-filled mine, the McKenzie mine that was not marked on any Saxsewell mine map. Mine director Workman added, "Some of the underground maps must have been wrong. According to the Gauley Coal and Coke Company chart, the crew was staying 200 feet away from the property line as the law requires. Either the old Straley map (another old name of the abandoned mine) or the Gauley map must have been wrong."[6]

Rescuers made an attempt to try and reach the missing ten through this mine. Every tunnel was completely filled to the top. Additional pumps were installed and the effort to help those still out of contact continued.

The third day, Wednesday, began with the six men deep in the mine still huddled on the ledge. Trying their best to stay on a normal flow of time, they continued to check the water levels. Moore wrote in his notebook: *5-8-68 Water went down about 1 ½ inches on day shift & 2nd shift about all we did was check water and prayed that someone would get to us.*[7]

Conditions for the 15 men at higher elevation were only slightly better. It was a cold and damp cramped space where only a few could stand at one time. The main difference was that food was adequate and with communication to officials at the surface intact, the men were optimistic that rescue was just a matter of time.

As the sun was setting Wednesday evening, rumors circulated that rescue of the 15 was imminent.[8] Nearly 500 people gathered including ambulances, news reporters; and crews from the major television networks with anticipation of a rescue. But it was not to be. Despite the pumping operation, the relentless water remained at dangerous levels not affording any rescue personnel a path down to the men.

Finally, the crowd dwindled except for anxious family members. Inadvertently, officials left the families waiting in the dark. Frustration began to mount as it would in any situation filled with so much uncertainty and fear, and, in this case, life and death.

"I don't believe what they say anymore," said the wife of one of the trapped 15. "On Monday they said it would be 24 hours, yesterday they said it would be 24 hours. Now they say it will be another 24 hours."[9]

"I don't know what's going on," another woman commented. "And I don't think those men down there know."[10]

In attempting to give the miners' families information on rescue timelines, a false sense of trying to reassure them came before any definite plans had been finalized. The main reason it was difficult for officials and rescue teams to gauge how long it would take to get to the men was the water. Pumps and teams of men were working feverously around the clock trying to see significant drops in the water levels. It was not happening.

Mike Davis said his sister Dinah recalled her father was trying to calculate when they would get out based on how much water was being pumped out. "Dad told her later that he became very frustrated because they would not tell him the pump breakdowns." Davis said. [11]

Dark Days

Families of the ten still unaccounted for had to anxiously wait and watch as rescue workers and volunteers were focusing on getting the first 15 men to the surface and safe. With no significant changes in the water level measurements in the area where Moore and his crew were thought to be, officials and the media concentrated on the rescue. Word on the "lost ten" would come when more accurate level readings allowed further decisions on how to proceed deeper in the mine.

The six entrapped men tried to stay focused on what they could control; which wasn't much. They heard drilling and were positive they would be in contact with their friends soon. They let themselves believe it wouldn't be too much longer due to the fact that they were running out of food. They all began to lean on their faith. Larry Lynch was the most open about his faith in God and he prayed for the group before they ate.[1] A sandwich was cut into six pieces and shared equally. What little food they had held out until the next day. Each man slowly consumed what made up the final morsels they would have for a while.

It was now Thursday. Just over 72 hours had passed since they went on shift when they finally exhausted their very small food and fresh water supply. The men were forced to drink mine water from that point on.[2]

John Moore Jr. noted: "May 9, Water still the same, not down any from yesterday···We have been here 72 hours. Scarbro and Gene Martin checked on water to see if is down or not. Water about 2 inches high[er than] it was. They started drilling a hole down top at 11:55 a.m. Hole drilled through at 3:55 p.m. Drilled through about 85 feet from us but we sure was glad to know they were working to get us out. Water level still the same. Scarbro and Fitzwater and Martin was beating on pan line trying to get them to hear outside. We thought sure they would drill another hole through to us. So we eat what we had left, which was 1 sandwich, 1 pt. of water.[3]

Vicki Rose said her father had concerns when they heard drilling, but he didn't convey that to the other men. "My dad felt responsible for them. He told us after he got home that he never said anything to the others, but he worried that if they drilled near their space which he thought was an air pocket, the water would be forced in and they wouldn't have had anywhere to go. He was glad they were still trying to get to them, but the drilling made him nervous, for sure."[4]

The air quality in the mine remained good, but they did struggle with coldness. They could not get comfortable. Chilly temperatures, dampness, and no physical activity make sleeping more and more difficult. They talked about God, their favorite foods, the best country-western singers, and did what they could to pass the time.[5] One thing they had in abundance was darkness. It did not help matters. They were all wearing helmets with headlamps; they had the foresight to conserve the use of their lights from the outset of the ordeal. There was great fear, even though none of the men would admit it, of being completely left in the dark. With their headlamps off and submerged within a small pocket, surrounded by nothing but water and darkness, they could not see the hand in front of their face. With their minds racing, sleep was hard to come by.

Vicki and her mother huddled with the others and waited for word. "Every once in a while someone from the mine would come over to the tent area and give my mom and the others an update. "It was always about the levels of the water. If the water did not go down, they couldn't really do anything. I just kept staring at the mine opening, visualizing my dad raising his head up out of that hole and running over to us."[6]

Another rescue deadline for the other 15 had come and gone. Mine officials now pinpointed Friday evening at 6 p.m. This time, they were certain. Hope among management for rescue of the "lost 10" took a hit. A hole they had drilled was thought to have been directly into the chamber where they were believed to be located. The hole indicated the chamber was completely filled with water. The trapped men heard rescue workers calling through the hole. They shouted back. But nobody heard them. "We believe that whole area is flooded," said Rescue Supervisor H.E. Sundstrom. "We are not drilling any more holes there."[7] (Later it was realized since the maps had not been updated, officials were drilling in areas a short distance from where the miners had actually been working.)

The six miners huddled in silence back to back with each other trying to stay warm. They hugged the heavy, motionless conveyor belt trying to squeeze out any warmth from the rubber fibers, anything to shield them from the harsh surroundings. Rose said her mother never stopped giving up hope. "She just knew dad would make it. Even when the papers and some of the folks that were waiting started to bring up that they may not get all of them out, she never doubted he would be okay. She had faith that he would make it out of there and back to her."[8]

Week from Hell

Friday morning came and went. The men had been trapped for an entire week on their ledge having gone four days and nights without any contact with the outside world. Hope that they would be rescued waivered back and forth along with the rare stomach pang after consuming nothing but mine water for almost a week. They were starting to lose weight, their bodies now getting nourishment from within to keep going. Jennings Lilly tried to keep things lightened up by half joking that the company would at least come after the valuable mine equipment that would not have been too badly damaged after a few days submerged in water.[1] But the serious side to their situation remained. The men could not figure out why they did not hear anyone trying to find them.

Continuous pumping around the clock had lowered the water for the ones trapped at the mile mark of the main entry. Davis and Rudd finally were able to join the other 13.[2] The mine entrance was still blocked by a "swag" - a floor to ceiling trough of water roughly 200 feet long.[3] Scuba divers had been called in early on in the week, but their diving equipment was way too bulky for the narrow and very cramped mine passages for them to get to the miners. The water levels had finally dropped to a point that officials could put a solid timetable on when they could get to the men. It would now be just a matter of hours until the 15 could be reached and rescued.[4]

Newspapers reports informed anxious readers about the impending rescue and reunion for wives who mused about all the overtime their husbands had earned while being trapped. One publication included an aerial view of a map looking straight down on the mine showing where the 15 were located. They also diagrammed the location of the ten who were believed to be dead.

Vicki and her mother had been routinely coming to the mine since Monday evening. From early in the morning until dark, they waited for any word. A very long and nervous week was coming to a close. So far it was always the same update, officials were only in contact with the miners from the group of 15. No news on any of the other ten. They would know more when the water levels went down. That was always the answer to any questions from worried family members. It was becoming clearer that most thought that once the 15 men in the higher elevation of the mine were out, it would become a recovery operation. "I think people didn't want us to have a false hope. It didn't matter what anyone said to me, I wouldn't believe anything had happened to my dad until there was proof," Vicki Rose said.[5]

Moore: May 10··· Water raised about 6 inches. It doesn't seem like they are trying very hard to get us out. We wonder why they didn't drill another hole through the top to drop us something to eat and drink. Time now is 3 p.m. Friday. We all try to sleep Friday night. We were all about dried out by now.[6]

The ordeal was about to end for the 15 miners stranded but never out of touch with the surface. Word began to circulate that the rescue operation to bring the men out could happen at any minute. Ambulances were standing by ready with medical personnel on call to administer any first-aid that may be needed. Other miners from the Saxsewell No. 8 Mine and miners from other companies in the area came to offer support and see if they could help in any way. Miners from other states even felt compelled to make their way to West Virginia and provide any assistance that may be needed. It didn't matter if they were union or non-union, their mining brothers were in trouble and they were there to see them make it out. Some had been there since they first heard about the accident. By now the nation was watching and the group of people gathered at the opening had grown. Members of the local and national media were well represented. Some had filed their stories and reporters did on camera reports about the imminent rescue. The national networks hoped for a "live" rescue during their evening time slot. But officials and rescue workers determined again that the water levels were still not safe yet to proceed.

The media and a national audience would have to wait as the rescue didn't happen in time for Friday's evening news. Vicki and her mom drove home again as evening came on. "Even though she tried not to show it, I could always see the worry and sadness on my mom's face as we left the mine each evening." Vicki Rose said. "She just didn't want to leave without him. They were high school sweethearts and had always been together. Not knowing if he was okay was very difficult on her."[7]

Daylight

Although the 15 men waiting for rescue were dealing with discomfort, they did have the peace of mind that it would now just be a matter of time that they were going home. Having remained in contact with the surface from the outset of their ordeal, their spirits were good. Southern Mines General Manager Tim Salvati had food sent to the first group of 13 men before the drilling of any air holes were complete. He placed food staples, water and coffee inside six, three-foot waterproof plastic tubes, wrapped them securely and sent them in on the conveyor belt.[1] Since midday Tuesday, May 7th the men had eaten regularly and their morale was good. But nearly a week of confinement was not without trials. Dr. Lee B. Todd of Quinwood was called when one miner, Ottie Dillon became ill. Dr. Todd had antacids sent in with instructions over the radio-telephone to treat Dillon's active stomach ulcer. "Most of the boys down there have been my patient at one time or another," said Dr. Todd. "In fact, I brought some of them into the world." "I've watched them grow up and go down that mine."[2]

The water line had reached just 300 feet from the mine entrance right after the initial flood Monday morning. After continuous pumping, the water level dropped to 800 feet from the mouth by Wednesday evening. Now, in the pre-dawn hours Saturday morning, officials were certain the water levels had dropped enough to begin the rescue operation and bring the men to the surface. The pumps that had been working around the clock were shut off and silenced while the 15 men cautiously waded alongside rescuers through a foot of water. They were led to transfer vehicles and then were placed on the conveyor belt for the ride out. [3]

The 200 plus family and friends waited quietly, anxiously until 5:20 a.m. when the first miner, Elwood O'Dell, emerged from the mine. He was weak and had to be helped from the 26-inch wide conveyor belt normally used to carry coal. Laughter and cheering followed as one stiff and chilled miner quickly followed another off the conveyor belt they had ridden into the bright floodlights at the mine entrance and freedom.[4]

After the first miner was helped to a waiting foam covered bench to be checked out, it only took five minutes for all 15 men to make it out of the mine. They had been trapped for 118 hours and 20 minutes since reporting to work the previous Monday morning.[5] They appeared somewhat dazed from the lights and all the attention, but all were in overall good health despite their confinement for just under a week.

"This is really it," screamed Mrs. Lonnie Bennett, "this is really the best time." Her husband's wide grin splitting his blackened face having escaped his brush with death.[6]

H. E. Sundstom, an official with the Maust Coal Company, the parent firm of Gauley Coal and Coke, said rescuers entered the mine at 3 a.m., "and made contact with the miners" 15 minutes later.[7]

The miners resembled ghosts when they emerged from the shaft into the bright light. They were thickly bearded, their faces hardly distinguishable under a thick layer of coal dust and dirt. "I'm fine, I'm fine," Glen Amick told his family as he walked to the ambulance, shaking hands with well-wishers along the way. Ottie Junior Walton, when asked if he would go back into the mines after his close call with death said, "Mining is the only thing I know. Someone has to pay the rent."[8]

The men had earlier agreed over the radio to be checked out by Dr. Lee B. Todd, the mine physician and taken to hospitals at Summersville and Richwood for observation. Those plans slightly changed as most of the men were in good shape and Dr. Todd determined despite the cramped conditions they endured, there was no need for the most of them to go to the hospital. They were allowed to go home by ambulances; their relatives followed.[9]

Nine miners went to a nearby hospital in Richwood for examination.

Frank Davis was one of the men taken to the hospital after the rescue. "My dad was placed in an ambulance for precautions, Mike Davis said." "I rode in the front and the driver was a man we knew named Fred Painter. He was driving over the mountain dirt roads very fast. My dad told him to slow down and Painter replied, 'I want you to be the first to arrive at the hospital!'" Davis was released from the hospital Sunday, May 12th, eager to get back to the mine. [10]

"They are strong men and all appear to be none worse for their experience," said Sister Mary Monica, administrator of the Sacred Heart Hospital. "Some are suffering from mild dizziness and a weakness in their legs, but this is to be expected."[11]

Roy Lee McClure was another one of the miners that went to Sacred Heart Hospital for a medical checkup and rest. McClure, president of the United Mine Workers Local 1254 of Richwood, would be fine and he was thrilled after visits from his two Air Force sons. The two had journeyed a total of 18,000 miles and their grueling trips ended happy with the reunion. S. Sgt. Roy Lee McClure Jr., 28 at the time, traveled 54 hours from Clark Field in the Philippines to be with his father after the rescue. Sgt. Paul McClure, then 21, traveled nearly 60 hours from Takhli, Thailand. The brothers were given emergency leaves and priority transportation from the Air Force after notification that their father was one of the 25 men in the mine at Hominy Falls.[12]

"It was a rough trip," Said Roy Jr. "My brother

and I were erroneously notified by the Red Cross that 'all hope had been abandoned' for the twenty five trapped men. "It was only when I reached Travis Air Force Base in California that I was told hope had only been given up for ten of the men, that our father was among the other fifteen."13

The elder McClure had been a miner for 19 years. "I would like to finish my 20," He said. "That way I won't lose any benefits."14

Within a few minutes, after days and days and false announcements of rescue operations it was all over. With the sun rising over the hilltop Saturday morning, May 11th, the mine entrance location now carried an eerie calm and quiet around it. Most of the throng of people who witnessed the 15 miners emerge just a couple hours earlier had left. Jubilant families of the rescued filed out, glowing with excitement, reunion, and answered prayers. There were very few people from the media still there. Their stories and reports of miners rescued were written, recorded, and on the way to television stations and newspapers around the country. Early edition newspapers carried the story with photos of weary, blackened-face miners smiling and hugging.

Their entrapment inside the flooded mountain was over. The ten still entombed in the mine were mentioned and noted in articles as – lost.

For a week, front pages around the country carried headlines with the story of an imminent rescue in the dangerous and confined underground world of coal miners. The public was anxious for information on how these men coped while being trapped inside a mine and not knowing when and how they would be rescued. The story of recovery did not have the same appeal as a daring and courageous rescue. Interest around the country about doomed coal miners in West Virginia went away as quickly as the mine itself had been compromised by the inundation of water. The news around the country and coming out of Vietnam during the week following the rescue of the 15 miners was very disturbing, and carried very high interest. It was the worst week for the United States military fighting in Southeast Asia up to that point. 562 Americans were killed and another 2,225 were wounded battling the Communists spring offensive.[15] Also that week, a devastating series of tornadoes ripped through ten states from Minnesota to Arkansas, killing more than

81

70 people. The twisters killed 49 in Arkansas alone and injured more than 500.[16]

The presidential election campaigns were taking candidates around the country on stops. President Lyndon Johnson's early withdrawal from the race that March put those in the running for the Democratic Party nomination under the microscope. Political rallies for candidates Robert Kennedy, Hubert H. Humphrey, and Eugene McCarthy bidding for their party's nomination were followed with keen interest. And NASA continued America's race to get to the moon detailing plans for manned space flight into orbit later that year. These stories now gathered all the attention.

The general public outside of this small West Virginia hamlet cared very little about the recovery operation to find dead coal miners. They had heard and read about that sad story from Appalachia many times before.

Families of the ten still out of contact were not aware of what was going on around the country, in Vietnam, or outer space. Wives, parents, and children of the miners feared lost huddled together.

They somehow hoped to find comfort in numbers. They faced new challenges. There was no longer a connection to the underground. Any hope that the 15 men rescued could shed light on their loved ones vanished once the men confirmed that they were cut off from the others. The miners had nothing to add to what mine officials had been telling the families all along. There were far fewer media members now present and families had fewer people around to find out information.

As bleak as it did look, the rescue operation did continue. Officials gave updates to the families as best as they could. The weekend began with the sound of pumps back on and churning again, racing to eliminate the water. "We will continue to drain the mine until we make contact with the remaining 10 men," said C.E. Richardson, president of the Gauley Coal and Coke Company.[17]

Footprints in the Mud

Huddled together, the "lost" group had no knowledge of the rescue. The only thing they did know was it had been two days since they heard any drilling. They were also aware that they had been out of touch long enough for many, even their families, to have probably given up hope.

And with no pumps operating, the hope of any rescue was fading. However, their worse fear was the fact that the water levels near their "tent" were actually rising. At about the time rescue officials were planning to begin the operation to go after the other 15, Moore wrote··· *5-11-68 Sat "At 1:40 a.m. everyone was awake and Scarbro and Lilly checked water about the same. John Moore checked water at 3:00 a.m. Sat. It has raised about 3 inches.*[1]

With little to do, the men waited for increments of time to pass and then measure the water again. They did anything to maintain hope and pick up their spirits. It was just human nature to make the best of their very dire situation. They tried to get comfortable moving around their tight quarters an inch at a time.

Moore: *We would lay down for a while and then sit up for a while every day and night. Sat. at 1:00 p.m. Joe checked water and it had raised about 8 inches more. We all prayed for it didn't look like they were pumping on it.*[2]

Moore and his wife Gladys were members of the Upper Buckeye Presbyterian Church. They were active in the church and went about every Sunday. The church did not have a choir, but the Moores including their daughter Vicki enjoyed singing. "I think my dad being a man of faith always held onto that while he was in the mines. He had comfort knowing that if anything ever happened to him he was ready and that he would be okay and God would take care of all of us," Vicki Rose said.[3]

Moore: *Sat at 5:00 p.m. I checked water it had went down about 1 inch.*[4]

As Vicki and her mom drove over to the mine Sunday morning, the congregation at their church in Buckeye gathered for morning service. The people came together to pray for the Moore family, and ask for John's safe return and for peace to come to all the miners involved and their families. It was

Mothers' Day and Vicki mentioned it to her mom on their drive. She knew it would not be a happy one for her mom; and she left it at that. It had been difficult for her mother to eat or sleep. That was common for the other wives too and many of the children. They had heard terrible rumors that the mine may eventually be sealed shut. They were reassured though by other miners who vowed to work until they found a way into the lost miners and they were found, dead or alive.

Inside the mine, all the men could do was watch the water levels and deduct if it would ever be safe enough to try and make it out on their own.

Moore: *5-12-68 Sunday morning at 5:00 a.m. Checked water. It had went down about 1 inch in about 12 hrs. It looks like they could do better. 1:00 p.m. J. Lilly and Scarboro checked water. It had went down about 1 inch. It is going down about 1 inch and hr. now. They are doing good. I went to check on the water at 4:30 p.m. It was down about 7 inches. We all prayed. At 7:30 p.m. Scarboro went to check on water. It was down 3 inches more.*[5]

An adjacent mine company, S &C Coal, had

been drilling holes through its own mine through the abandoned McKenzie Mine. They had been pumping water out and many of the holes they had made had run dry. They decided to make a big opening and get to the men by blasting away the coal.[6]

It had been eight days since the men had anything substantial to eat. Four days had passed since they shared the last morsels of a sandwich. Now it was nothing but acidic mine water. Moore's usually neat handwriting became difficult to read. He chronicled the group's limited activities along with blotches and scribbles in his notes.

Moore: *5-14-68 Monday. Water dropping all the time. We all thought we could make it outside. So we tried but when we got to the swag we couldn't get across it. Water was still up to roof. Joe Fitzwater passed out and couldn't make it back. We put him on a shovel and pulled him back. I guess we were in a little black damp. But I couldn't get my safety light to light. I got water in it. Monday about 3:30 p.m. Went to see if could find a telephone. Gene Martin went with me to get water. I couldn't find the telephone. It had washed away. I found the first aid box and started back with it.*

When I cross the belt I found one of the men from the #3 miner laying beside belt···I had to cry. I didn't have the nerve to look to see which one it was. Gene and I made it back with the others. I soon got hold of myself and dressed Gene Martin's hand.[7]

Moore's notes of the water dropping were accurate. Pumps were finally able to be used in Moore's section after the entrance was cleared when they rescued the first group. The S&C miners completed a bigger opening through the McKenzie Mine. They faced other obstacles though as the opening carried black damp (a dangerous and potentially deadly oxygen deficient air). The area was ventilated with fans and readings were taken. But the path in was blocked by roof falls.[8] The miners were not deterred and worked non-stop to clear the way further into the mine. They were making progress and sensed they were getting close to finding the men.

All Moore and his men had was the fetid mine water. They prayed to God that the water would be cleansed and they could drink. Jennings Lilly reminded the others how important it was to continue to try and drink the water. Even though they didn't have anything to eat, they had to continue to drink the polluted water to stay alive.

5-14-68 Tuesday morning. Everything seems pretty quiet. I guess we will just sit here and wait to see if anyone comes to get us. It seems like a long time we have been here. I think I have eat[en] about all the skin off my lips. We have all been praying. Lynch sure has been a big help to all of us. He has prayed about all the time.[9]

Frank Davis had returned to the mine after regaining his strength determined to get in there and get his guys out. Company Manager T.A. Salvati had been working on the rescue since the beginning. Davis and Salvati tried to work their way into where the men had been mining coal the previous Monday morning. But Salvati told the eager Davis it was still too dangerous and the water levels would still not permit them to move closer. The men worked around the clock and rescue personal relieved one another on rotating shifts every few hours.

Davis and Salvati and the rescue teams worked as quickly as they could to penetrate the murky passage and find any signs of the men. They had not yet made it to the Number 3 Room working area where the flood had started. But they were close.

5-15-68 Wed. morning at 1:30 a.m. Everything quiet. I dressed Gene Martin's hand. He has a pretty bad hand. Wed. 7:30 a.m. Gene Martin and Scarbro went after water. Scarbro passed out. We had quite a time getting him back to what I called the tent. I think he drink too much water. 8:30 a.m. Everyone settled down. But Scarboro still a pretty sick man. 2:30 p.m. We set up a while and talked. Water still going down. I feel pretty good now. I am pretty sure we will get out alive. Thank God. 8 p.m. All of us are going to try & go to sleep awhile.[10]

Rescue workers were in *rescue* mode, but most thought there would not be any way they would find any of the ten men alive. They had witnessed the devastation the powerful flooding had caused along the passageway to where the men were supposed to have been working. The water level had continued to recede and they moved sloshing along slowly through the cramped tunnels as they had done now for the last three days. This time though it was different. Jim Blankenship of the United States Bureau of Mines was wading along through knee high water in the Number 1 entry off two right main tunnel. His headlamp caught something familiar in the mud just above the water line -footprints! And they looked fresh.[11]

The rescue group also included Clyde Perry and Fred Castell, of the United States Bureau of Mines, and Kermit Stanley a nearby coal mine owner. They continued on another 650 feet. As a group, they decided it was best to return to the rescue base with this new and surprising information.[12]

Moore: At about 1:10a.m. we heard men talking. They are coming after us.[13]

Now the additional crew, better equipped for rescue, began the descent into the mine with renewed enthusiasm. An hour had passed since the previous group had found the footprints; the rescue crew made their way toward the zone where the men mostly likely would be located. They slowly moved along another twenty yards when the rescuers first noticed some debris on the side of a slight elevated ledge. Looking up their lights aided their first visual contact with the makeshift brattice curtains the men had constructed in a panic ten days before.

Moore: We hollered at them. Roy Simmons was the first to put his head through to us. I think we all hug him. We sure thank God for saving us.[14]

Upon reaching the men, Blankenship and the others were both amazed and excited that they found men still alive. They saw tears in Moore's eyes. "At first the men didn't say anything. Then after a few moments they ask about the other 15 men on their shift. Then they asked about their families. Finally Moore said they were ready to go, Blankenship said.[15]

Rescuers gingerly helped the six exhausted men make their way from the ledge in their Number 2 Room shelter through 1,000 feet of water and mud. At a waiting mantrip location, they were helped into transfer cars for the long awaited trip out of the mine.[16]

The men finally reached the surface at 4:30 a.m. They were hardly recognizable. Their unshaven faces were covered with coal dust and mud. A small group had gathered during the night and some of the miners greeted the few loved ones that were there. The six had been cut off from the world for just shy of 238 hours.

"My mother and I were not there when the men came out," Vicki Rose said. "It was the middle of the night and not many of the families were there waiting. My mother got a call that they had rescued the men and that they were being taken to the hospital."[17]

Ambulances transported the men to the Sacred Heart Hospital in Richwood. They were found to be in overall good physical condition. They were given hot cereal, toast, and milk. The men stayed in the hospital for varied short stays for rest and different treatments.[18] "I can just remember how elated my mom was and how matter of fact my dad seemed," Vicki Rose said. "He told her he didn't know what she was so worried about, he knew he was fine and he was going to make it out all along. He always liked to kid with her and he was so thrilled to be back with her. He looked so skinny. He never was very big, but you could tell he had lost weight. And he had that eye makeup (as I called it) still on from the coal dust. He laughed and joked, but cried when we hugged him. He was so thankful to be out and well. I really could feel how much they loved each other and it was just wonderful to see my parents reunited."[19]

93

The miners had different opinions on how the ten days trapped in the mine affected their futures of going back to work so far underground every day. "I expect I'll go back." John Moore said. "I've been a coal miner 19 years. These things happen," Moore said right after the rescue.[20]

Larry Lynch, who was 28 when he was one of those feared dead, said he would go, "wherever Gods calls me. If I can serve Him best down in that mine I'll go down there."[21]

The bodies of the four men who died upon the inrush of water were located and identified. Cause of death was drowning.[22] The deceased were brought to the surface at 8:30 a.m. Saturday, May 18th, 1968.[23] Those who perished were: William Frank Burdette, timberman, 43 years old from Rainelle, West Virginia. He had 14 years of coal mining experience. Burdette was survived by his wife Rowena and two children. Claude Roy Dodd Jr., a miner's helper, was 42 years old. He also lived in Rainelle and had worked in the mines for 24 years. Dodd was survived by his wife Arlene and two children.

Timberman Renick F. McClung was 46 years old. He lived in Orient Hill, West Virginia and had been a coal miner for 27 years. McClung was survived by his wife Helen and five children. Eli Edward Walkup was a miner operator and died at age 37. He lived in McRoss, West Virginia and was survived by his wife Hilda and three children.[24] Eli and Hilda Walkup's fourth child, Shelly, was born June 6th, 1968. Exactly one month after the accident happened.

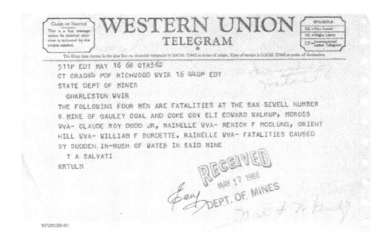

Gauley Coal and Coke Company telegram
to the West Virginia Department of Mines.
Courtesy West Virginia state archives.

Overall, the six West Virginia men that survived, Moore, Lynch from Richwood, Joe E. Fitzwater, 33 of Rupert, Jennings H. Lilly, 30 of Mt. Nebo, Edward F. Scarboro 38, of Richwood, and Eugene H. Martin, 31 of Clintonville, believed they survived on the small ledge for so long due to divine intervention. It came down to the water they drank to stay alive. The men felt it was a miracle when they prayed and asked God to make the dirty mine water clean enough to drink. The water was then clean and they drank.

Fitzwater said it was a miracle they made it out. He believes they survived thanks to his friend, Larry Lynch. Lynch prayed to God every day to keep them alive. "It's a miracle because we were in there for ten days without a bite of food. The floodwater sealed all of the places off. We didn't have any air. He prayed for air, we got air. He prayed for food, God put food in that water because everything got washed away. I can only speak for myself, but I drank it three times a day. I never had a hunger pain in my stomach," said Fitzwater. Fitzwater said the event changed his life for the better and led him to the Lord. "I just believe Him like brother Larry believed in Him. God is just as big as you want Him to be," says Fitzwater.[25]

"I went in a sinner and came out a Christian," Fitzwater says. "It was the best thing that ever happened to me. Each day we grew weaker, but we kept our heads. We survived through the grace of God. It was a miracle of God's."[26]

Ed Scarboro never mined coal again. He relocated to Michigan after the accident and went into the construction business. He returned to West Virginia to live in 1986. Scarboro said he had gone through a lot of bad situations in his life, but surviving the Hominy Falls mine flood was the toughest. "It was ten days of hell. Some of them guys broke down and cried just like they were little kids. I'd never let a man see me like that, but I was just as scared as they were."[27]

Gene Martin's hand healed and after an extended time off, he returned to work. The scar on his hand is a constant reminder of how blessed he was to have survived. "At no time did any man try and leave and make it on his own." Martin said. He credits Lynch, the part-time minister, for helping them make it through the ordeal. "He (Lynch) was the one to lead us.[28]

"It changed my life completely. I became a Christian after that. I realized back there inside that mine that I wasn't satisfied with my life the way I had been living it," Martin said. "Sometimes, I wonder why I was the one who survived."[29]

"I fulfilled a vow to God. I told Him when I got outside that I would thank Him and give Him praise for delivering us sound and alive. Not just me but the other five men who were with me," Lynch said.[30]

Jennings Lilly lost 30 pounds during the ordeal from his 170 pound frame. "I told those boys, 'a man can live a long time without food, but not water, Lilly said, and that's what saved us."[31]

Lilly was the miner operator working the No. 1 room that morning. He did not know he was cutting within 5 to 7 feet on a parallel course along the water-filled Sugar Grove No. 4 old timer for some feet before turning his miner.[32] He had been making short cuts because there was more rock than coal at that point. Had he made his customary wide cuts, there is no doubt he would have cut through into the old mine before Walkup did. "I am convinced it was God's will that his group was saved and His will that four of their friends and buddies lost their lives," Lilly said. [33]

The rescue was of great relief to Lilly's son who was eight years old at the time. He had gone on a "strike," refusing to go to school until his dad was safe. [34]

There were many others in the community that refused to give up on those considered lost. The Salvation Army had set up a truck and the Red Cross had a mobile unit on location for the duration of the rescue. The Leivasy Boy Scouts were excused from school to help hand out sandwiches and coffee or run errands.[35] No. 8 mine was near the Mt. Urim Baptist church at Green Valley. [36] This church and others in the area provided food and blankets from the onset to families, rescue workers and media members.

Maust Coal and Coke President Chuck Richardson called daily press conferences about the status of the trapped miners. Southern Mines

Superintendent Tim Salvati did everything in his power to keep communication lines open to the trapped miners, families, officials and the media.

There were others who worked long, grueling hours to see the rescue through. Alfred Kincaid of Fenwick Mountain was a supply truck driver. He worked 53 straight hours at one point. Others who would not pull themselves away from duty were lampmen James Little and Paul Meadows of Richwood.[37] They provided order by making sure everyone knew where everything was located inside and out at the mine site.

Officials called on retired miner Denver Short for help. Short had experience running certain mining equipment in the rescue effort that no longer was being used in operations at the Saxsewell No.8 mine.[38]

Gerald Rader of Rader Flying Service often flew parts in for critical pump repairs from Charleston, (W.Va.) and Clarksburg, (W.Va.).[39]

The responders going underground in the rescue effort faced tremendous danger and obstacles but never wavered in their effort to get to the men.

They pushed their way through the darkness with just enough room to keep their faces from scraping the tunnel roof and their mouths and noses out of the contaminated water.

Rescue workers Roy Simmons and his friend Lowell Snyder went to the mine together every day from Cowen, West Virginia.[40] They were among the group including Pat Bennett and Jack Lee Mauls who finally reached the miners and help them get back to the where they had entered the mine ten days before.[41]

On May 18, 1968, The Gauley Coal and Coke Company sealed the hole between the active and abandoned mines. On June 12, 1968, mining in Saxsewell No. 8 began again.[42]

Mine Superintendent Frank Davis was considered a hero by many following the Hominy Falls rescue for his relentless efforts in finding the men after so many had given up hope. Following his service in the Navy during World War II, Davis' career as a West Virginia coal miner spanned 43 years. He died in 2013 at the age of 89.

Frank Davis in front of his home - 1961

After the rescue, John Moore, Jr. stayed in the hospital a few days longer than the other five men. His nephew came all the way from California to visit and his mother and sister came to see him from Baltimore. He had just got over a bout with the flu prior to going to work on May 6th. He had lost some weight and the extended stay in the cold and wet cramped conditions had greatly affected the arthritis in his back. He tried to return to work a couple weeks after the rescue, but the pain was too great and he decided to put in for retirement.[43] The coal company (Island Creek Coal Company took over the mine) paid him half his wages for the rest of the year. He was never given his retirement.[44]

Moore had a case for workman's compensation and won full disability after three years. He never went back to work in the mines again. He stayed active in his church, took much pride in his garden and remained content working in and around his home, always finding a way to take care of his family.[45] John Moore Jr. died in 1990 at the age of 68.

John Moore Jr.'s Handwritten Journal

Day 1, Monday, May 6, 1968.

Days 2 and 3, Tuesday, May 7 and
Wednesday, May 8, 1968.

Day 4, Thursday, May 9, 1968.

Days 5 and 6, Friday, May 10 and
Saturday, May 11, 1968.

Day 7, Sunday, May 12, 1968.

Day 8, Monday, May 13, 1968.

Day 9, Tuesday, May 14, 1968.

Day 10, Wednesday, May 15, 1968.

Day 11, Thursday, May 16, 1968.

Notes:

Chapter 1 Miner's Morning

1. Interview with Vicki Rose, July 2015.
2. Interview with Vicki Rose, July 2015.
3. www.weatherunderground.com/history
4. United States Mine Rescue Association. http://USminedisasters.com/saxsewell-mine – conditions immediately prior to inundation.
5. United States Department of Labor. Bureau of Labor statistics. http://data.bls.gov/pdq/surveyoutletservlet
6. United States Department of Defense. http://thewall-USA.com/names.asp
7. National Vietnam Veterans Foundation.org
8. United States Department of Defense. http://thewall-USA.com/names.asp
9. Interview with Vicki Rose, July 2015.

   Vietnam Conflict Extract Data File, as of April 29, 2008, of the Defense Casualty Analysis System (DCAS). Part of record group 330: Records of the Office of the Secretary of Defense. http://aad.archives.gov/aad/record-detail

   http://www.archives.gov/research/military/Vietnam-War/casualty-lists/WV-alpha

10. Associated Press. (Multiple newspapers) http://www.History.com/topics/Vietnam-War
11. NASA.gov-President Kennedy Speech to Congress May 25, 1961. http://www.space.com/11772-president-kennedy-historic-speech-moon-space.html

Chapter 2 Mantrip

1. West Virginia Office of Miners' Health, Safety and Training. West Virginia Geologic and Economic Survey. Green Lands published by the West Virginia Mining and Reclamation Association. Charleston Gazette-Mail. National Mining Association, Impact: The Importance of Coal to West Virginia published by the West Virginia Mining and Reclamation Association.
2. Fatal Accident Report-Hominy Falls Mine Disaster. "Description." West Virginia Division of Culture and History. http://www.wvculture.org/history/disasters/hominyfalls02.html
3. John Moore, Jr. testimony, Hearing on Hominy Falls mine disaster. West Virginia state capitol, Room 410, Charleston, West Virginia, May 24th, 1968. West Virginia Division of Culture and History.

4. United States Mine Rescue Association. http://USminedisasters.com/saxsewell-mine Evidence of Activities and Story of Inundation.

5. Fatal Accident Report-Hominy Falls Mine Disaster. "Description." West Virginia Division of Culture and History. http://www.wvculture.org/history/disasters/hominyfalls02.html

6. United States Mine Rescue Association. http://USminedisasters.com/saxsewell-mine Evidence of Activities and Story of Inundation.

7. Interview with Vicki Rose, July 2015.

8. United States Department of Labor. http://www.dol.gov/whd/minwage/chart.htm

9. United States Mine Rescue Association. http://USminedisasters.com/saxsewell-mine Mine conditions immediately prior to Inundation.

10. United States Mine Rescue Association. http://USminedisasters.com/saxsewell-mine Mine conditions immediately prior to Inundation.

11. United States Mine Rescue Association. http://USminedisasters.com/saxsewell-mine Mine conditions immediately prior to Inundation.

12. West Virginia Mine Disasters 1884 to Present. http://www.wvminesafety.org/disaster.htm

13. West Virginia Mine Disasters 1884 to Present.
    http://www.wvminesafety.org/disaster.htm
14. Davitt McAteer, Monongah. West Virginia University Press, copyright 2007.
15. West Virginia Mine Disasters 1884 to Present.
    http://www.wvminesafety.org/disaster.htm
16. John Moore, Jr. testimony, Hearing on Hominy Falls mine disaster. West Virginia state capitol, Room 410, Charleston, West Virginia, May 24th, 1968. West Virginia Division of Culture and History.
17. United States Mine Rescue Association. http://USminedisasters.com/saxsewell-mine Mine conditions immediately prior to Inundation.
18. United States Mine Rescue Association. http://USminedisasters.com/saxsewell-mine Mine conditions immediately prior to Inundation.
19. Fatal Accident Report-Hominy Falls Mine Disaster. "Cause" West Virginia Division of Culture and History.
    http://www.wvculture.org/history/disasters/hominyfalls02.html

20. Fatal Accident Report—Hominy Falls Mine Disaster. "Cause" West Virginia Division of Culture and History. http://www.wvculture.org/history/disasters/hominyfalls02.html
21. Fatal Accident Report—Hominy Falls Mine Disaster. "Description" West Virginia Division of Culture and History. http://www.wvculture.org/history/disasters/hominyfalls02.html

Chapter 3 We've Hit Water

1. United States Mine Rescue Association. http://USminedisasters.com/saxsewell-mine General Information.
2. United States Mine Rescue Association. http://USminedisasters.com/saxsewell-mine General Information. Fatal Accident Report—Hominy Falls Mine Disaster. "Introduction" West Virginia Division of Culture and History. http://www.wvculture.org/history/disasters/hominyfalls02.html
3. United States Mine Rescue Association. http://USminedisasters.com/saxsewell-mine General Information.
4. United States Mine Rescue Association. http://USminedisasters.com/saxsewell-mine Mine maps as factor in inundation.

5. West Virginia Office of Miners' Health, Safety and Training. West Virginia Mine Map Archives.
http://www.wvminesafety.org/minemaps.htm

6. United States Mine Rescue Association.
http://USminedisasters.com/saxsewell-mine
Mine maps as factor in inundation.

7. Fatal Accident Report-Hominy Falls Mine Disaster. "Description" West Virginia Division of Culture and History.
http://www.wvculture.org/history/disasters/hominyfalls02.html

8. Fatal Accident Report-Hominy Falls Mine Disaster. "Description" West Virginia Division of Culture and History.
http://www.wvculture.org/history/disasters/hominyfalls02.html

9. United States Mine Rescue Association.
http://USminedisasters.com/saxsewell-mine
Evidence of Activities and Story of Inundation.

10. United States Mine Rescue Association.
http://USminedisasters.com/saxsewell-mine
Evidence of Activities and Story of Inundation.

11. Interview with Vicki Rose. July 2015.

12. Fatal Accident Report-Hominy Falls Mine Disaster. "Introduction" West Virginia Division of Culture and History. http://www.wvculture.org/history/disasters/hominyfalls02.html
13. United States Mine Rescue Association. http://USminedisasters.com/saxsewell-mine Evidence of Activities and Story of Inundation.
14. Tara Tuckwiller, Before there were Quecreek and Sago, there was Hominy Falls. Charleston Gazette, April 20, 2006.
15. United States Mine Rescue Association. http://USminedisasters.com/saxsewell-mine Evidence of Activities and Story of Inundation.
16. Fatal Accident Report-Hominy Falls Mine Disaster. "Description" West Virginia Division of Culture and History. http://www.wvculture.org/history/disasters/hominyfalls02.html
17. John Moore, Jr. testimony, Hearing on Hominy Falls mine disaster. West Virginia state capitol, Room 410, Charleston, West Virginia, May 24th, 1968. West Virginia Division of Culture and History.

Chapter 4 Head Count

1. United States Mine Rescue Association. http://USminedisasters.com/saxsewell-mine Evidence of Activities and Story of Inundation.

2. John Moore, Jr. testimony, Hearing on Hominy Falls mine disaster. West Virginia state capitol, Room 410, Charleston, West Virginia, May 24th, 1968. West Virginia Division of Culture and History.

3. United States Mine Rescue Association. http://USminedisasters.com/saxsewell-mine Evidence of Activities and Story of Inundation.

4. John Moore Jr. Handwritten notes during entrapment. May 6, 1968.

5. John Moore Jr. Handwritten notes during entrapment. May 7, 1968.

6. United States Mine Rescue Association. http://USminedisasters.com/saxsewell-mine Evidence of Activities and Story of Inundation.

7. Holger Jensen. (Associated Press).Charleston Gazette, May 11, 1968. Charleston Daily Mail, May 11, 1968. Tara Tuckwiller, Before there were Quecreek and Sago, there was Hominy Falls. Charleston Gazette, April 20, 2006.

8. Holger Jensen. (Associated Press).Charleston Gazette, May 11, 1968. Charleston Daily Mail, May 11, 1968.

9. United States Mine Rescue Association. http://USminedisasters.com/saxsewell-mine Evidence of Activities and Story of Inundation.

10. John Christopher, Miracle at Hominy Falls. Man's Magazine, October 1968.

11. John Christopher, Miracle at Hominy Falls. Man's Magazine, October 1968.

12. United States Mine Rescue Association. http://USminedisasters.com/saxsewell-mine Evidence of Activities and Story of Inundation.

13. John Christopher, Miracle at Hominy Falls. Man's Magazine, October 1968.

14. United States Mine Rescue Association. http://USminedisasters.com/saxsewell-mine Evidence of Activities and Story of Inundation.

15. John Christopher, Miracle at Hominy Falls. Man's Magazine, October 1968.

16. John Christopher, Miracle at Hominy Falls. Man's Magazine, October 1968.

17. United States Mine Rescue Association. http://USminedisasters.com/saxsewell-mine Evidence of Activities and Story of Inundation.

18. John Christopher, Miracle at Hominy Falls. Man's Magazine, October 1968.

19. John Christopher, Miracle at Hominy Falls. Man's Magazine, October 1968.
20. John Christopher, Miracle at Hominy Falls. Man's Magazine, October 1968.
21. John Christopher, Miracle at Hominy Falls. Man's Magazine, October 1968.
22. Tara Tuckwiller, Before there were Quecreek and Sago, there was Hominy Falls. Charleston Gazette, April 20, 2006.
23. John Christopher, Miracle at Hominy Falls. Man's Magazine, October 1968.
24. Tara Tuckwiller, Before there were Quecreek and Sago, there was Hominy Falls. Charleston Gazette, April 20, 2006.
25. Tara Tuckwiller, Before there were Quecreek and Sago, there was Hominy Falls. Charleston Gazette, April 20, 2006.
26. James A. Haught, Wall of Water Traps 25 Miners in Nicholas. Charleston Gazette, May 7, 1968.
27. Interview with Vicki Rose, July 2015
28. Interview with Vicki Rose, July 2015
29. Interview with Vicki Rose, July 2015
30. Interview with Vicki Rose, July 2015
31. Interview with Mike Davis, April 2017
32. Interview with Mike Davis, April 2017

Chapter 5 The First Night

1. United States Mine Rescue Association. http://USminedisasters.com/saxsewell-mine Recovery Operations.
2. John Moore Jr. Handwritten notes during entrapment. May 7, 1968.
3. Tara Tuckwiller, Before there were Quecreek and Sago, there was Hominy Falls. Charleston Gazette, April 20, 2006.
4. Interview with Mike Davis, April 2017
5. Interview with Vicki Rose, July 2015
6. James A. Haught, Wall of Water Traps 25 Miners in Nicholas. Charleston Gazette, May 7, 1968.
7. John Moore Jr. Handwritten notes during entrapment. May 8, 1968.
8. Tara Tuckwiller, Before there were Quecreek and Sago, there was Hominy Falls. Charleston Gazette, April 20, 2006.
9. Tara Tuckwiller, Before there were Quecreek and Sago, there was Hominy Falls. Charleston Gazette, April 20, 2006.
10. Tara Tuckwiller, Before there were Quecreek and Sago, there was Hominy Falls. Charleston Gazette, April 20, 2006.
11. Interview with Mike Davis, April 2017

Chapter 6 Dark Days

1. Tara Tuckwiller, Before there were Quecreek and Sago, there was Hominy Falls. Charleston Gazette, April 20, 2006.
2. Tara Tuckwiller, Before there were Quecreek and Sago, there was Hominy Falls. Charleston Gazette, April 20, 2006.
3. John Moore Jr. Handwritten notes during entrapment. May 9, 1968.
4. Interview with Vicki Rose, July 2015.
5. Sean D. Hamill, Miners Given Up for Dead in 1968 Say Miracle Is Still Possible in Utah. New York Times, August 20, 2007.
6. Interview with Vicki Rose, July 2015.
7. Tara Tuckwiller, Before there were Quecreek and Sago, there was Hominy Falls. Charleston Gazette, April 20, 2006.
8. Interview with Vicki Rose, July 2015.

Chapter 7 Week from Hell

1. Sean D. Hamill, Miners Given Up for Dead in 1968 Say Miracle Is Still Possible in Utah. New York Times, August 20, 2007.
2. United States Mine Rescue Association. http://USminedisasters.com/saxsewell-mine Recovery Operations.

3. Tara Tuckwiller, Before there were Quecreek and Sago, there was Hominy Falls. Charleston Gazette, April 20, 2006.
4. Tara Tuckwiller, Before there were Quecreek and Sago, there was Hominy Falls. Charleston Gazette, April 20, 2006.
5. Interview with Vicki Rose, July 2015
6. John Moore Jr. Handwritten notes during entrapment. May 9, 1968
7. Interview with Vicki Rose, July 2015.

Chapter 8 Daylight

1. John Christopher, Miracle at Hominy Falls. Man's Magazine, October 1968.
2. Holger Jensen, Miner gets Ulcer treated by phone. (Associated Press) May 10, 1968
3. Edwin K. Wiles, Pumps and Drills Aid Mine Rescue. (United Press International), Watertown Times (New York), May 8, 1968
4. Holger Jensen, Search for 10 Men Goes On After 15 Miners Are Freed. (Associated Press) May 12, 1968.
5. Holger Jensen, 15 Rescued From Mine. (Associated Press) The Herald Statesman (Yonkers, New York) May 11, 1968

6. Holger Jensen, 15 Rescued From Mine. (Associated Press) The Herald Statesman (Yonkers, New York) May 11, 1968.

7. Holger Jensen, Search for 10 Men Goes On After 15 Miners Are Freed. (Associated Press) May 12, 1968.

8. Holger Jensen, I Expect···I'll Go Back. (Associated Press) The Herald Statesman (Yonkers, New York) May 17, 1968).

9. Edwin K. Wiles, 15 Miners, Trapped 5 Days, Ride to Safety. (United Press International), Watertown Times (New York), May 11, 1968

10. Interview with Mike Davis, April 2017

11. Holger Jensen, Search for 10 Men Goes On After 15 Miners Are Freed. (Associated Press) May 12, 1968

12. Happy Reunion, Spartanburg Herald-Journal (South Carolina). (Associated Press) May 12, 1968.

13. Pacific Stars & Stripes, May 14, 1968

14. Happy Reunion, Spartanburg Herald-Journal (South Carolina). (Associated Press) May 12, 1968

15. Jack Walsh, 562 GIs Killed in Week. (United Press International) Watertown Daily Times (Watertown, New York), May 16, 1968.

16. Robert Benjamin, 74 Dead, Toll Mounts, As Twisters Devastate Portions of 10 States. (United Press International), Union-Sun Journal (Lockport, New York), May 16, 1968.
17. Tara Tuckwiller, Before there were Quecreek and Sago, there was Hominy Falls. Charleston Gazette, April 20, 2006.

Chapter 9 Footprints in the Mud

1. John Moore Jr. Handwritten notes during entrapment. May 11, 1968.
2. John Moore Jr. Handwritten notes during entrapment. May 11, 1968.
3. Interview with Vicki Rose, July 2015
4. John Moore Jr. Handwritten notes during entrapment. May 11, 1968.
5. John Moore Jr. Handwritten notes during entrapment. May 12, 1968
6. Tara Tuckwiller, Before there were Quecreek and Sago, there was Hominy Falls. Charleston Gazette, April 20, 2006.
7. John Moore Jr. Handwritten notes during entrapment. May 13, 1968
8. Tara Tuckwiller, Before there were Quecreek and Sago, there was Hominy Falls. Charleston Gazette, April 20, 2006.

9. John Moore Jr. Handwritten notes during entrapment. May 14, 1968.
10. John Moore Jr. Handwritten notes during entrapment. May 15, 1968.
11. The Milwaukee Sentinel, wire services, May 17, 1968
12. The Milwaukee Sentinel, wire services, May 17, 1968.
13. John Moore Jr. Handwritten notes during entrapment. May 16, 1968.
14. John Moore Jr. Handwritten notes during entrapment. May 16, 1968.
15. The Milwaukee Sentinel, wire services, May 17, 1968.
16. The Milwaukee Sentinel, wire services, May 17, 1968.
17. Interview with Vicki Rose, July 2015.
18. Union-Sun Journal (Lockport, New York), (United Press International) 6 of 10 Men Found Alive in Coal Mine. May 16, 1968
19. Interview with Vicki Rose, July 2015.
20. Holger Jensen, I Expect…I'll Go Back. (Associated Press) The Herald Statesman (Yonkers, New York) May 17, 1968).

21. Holger Jensen, I Expect···I'll Go Back. (Associated Press) The Herald Statesman (Yonkers, New York) May 17, 1968).
22. Fatal Accident Report-Hominy Falls Mine Disaster. "Introduction" West Virginia Division of Culture and History. http://www.wvculture.org/history/disasters/hominyfalls02.html
23. United States Mine Rescue Association. http://USminedisasters.com/saxsewell-mine Recovery Operations.
24. Fatal Accident Report-Hominy Falls Mine Disaster. "Introduction" West Virginia Division of Culture and History. http://www.wvculture.org/history/disasters/hominyfalls02.html
25. Kathlynn Stone, Remembering the Hominy Falls Disaster. WVVA.com posted May 19, 2014
26. Reading Eagle (Reading, Pennsylvania). September 10, 1970.
27. Sean D. Hamill, Miners Given Up for Dead in 1968 Say Miracle Is Still Possible in Utah. New York Times, August 20, 2007.
28. Milwaukee Journal, wire services. May 17, 1968
29. David T. Sibray, "We've Hit Water," Hominy Falls, 25 years later. The Coal Chronicle, May 1993.

30. Sean D. Hamill, Miners Given Up for Dead in 1968 Say Miracle Is Still Possible in Utah. New York Times, August 20, 2007.
31. Sean D. Hamill, Miners Given Up for Dead in 1968 Say Miracle Is Still Possible in Utah. New York Times, August 20, 2007.
32. Sherman E. Drumm, Black Monday. Richwood W. Va. Public Library. 1968.
33. Sherman E. Drumm, Black Monday. Richwood W. Va. Public Library. 1968.
34. James Comstock, Miracle at Hominy Falls. Comstock Load, The West Virginia Hillbilly, May 25, 1968.
35. James Comstock, 25 years ago this month, A Terrible Mine Disaster. The West Virginia Hillbilly, May 6, 1993.
36. Mary Claire Johnson, The Tragedy of Hominy Falls Mine Disaster is Remembered. Nicholas County (W.Va.) News Leader, June 10, 1998.
37. Many Heroes at No.8 Will Never Be Recognized. Nicholas County (W.Va.) News Leader, May 22, 1968.
38. Interview with Mike Davis, April 2017.
39. Many Heroes at No.8 Will Never Be Recognized. Nicholas County (W.Va.) News Leader, May 22, 1968.
40. Interview with Mike Davis, April 2017.

41. Sherman E. Drumm, Black Monday. Richwood W. Va. Public Library. 1968.
42. David T. Sibray, "We've Hit Water" Hominy Falls, 25 years later. The Coal Chronicle, May 1993.
43. Interview with Vicki Rose, July 2015.
44. Tara Tuckwiller, Before there were Quecreek and Sago, there was Hominy Falls. Charleston Gazette, April 20, 2006.
45. Interview with Vicki Rose, July 2015.

INDEX

Page numbers in italics indicate graphic or photograph

Amick, Glen, ... 42, 77
Apollo 7, (NASA), ... 18
Apollo 8, (NASA), ... 18
Auto Trucks, ... 20
Baltimore & Ohio Railroad 20
Beam, Charles, ... 24
Bennett, Lonnie C., ... 42,77
Bennett, Pat, ... 101
Bess, Harry, ... 42
bituminous coal, ... 22
black damp, ... 88
Blankenship, Jim, ... 90,92
brattice cloth, ... 39,42
Burdette, Frank (William), ... 5,33,*36*,37,94
Burdette, Rowena, ... 94
Castell, Fred, ... 91
Casto, Isaac L., ... 43
Chevrolet Carry-All, ... 11
Christ, Jesus, ... 6

Collins, Eldon J., ... 42

Copen, Addison A., ... 42

Davis, Franklin, (Frank), ... 31,42,46-50,55,57
58,60,70,78,89,*102*

Davis, Mike, ... 55,56,60,63,78

Dillon, Oscar, ... 42

Dillon, Osmond L., (Ottie), ... 42,75

Dodd, Arlene, ... 94

Dodd, Claude Roy, Jr., ... 5,37,94

drift mine, ... 22

Fairmont Coal Company, ... 26

Femco radio phone, ... 47

Fitzwater, Ernest, ... 21,43-47

Fitzwater, Joe, ... 37,38,43,47,66,85,87,96,97

Gauley Coal and Coke Company, ... 13,14,20,30
31,43,47,53,61,77,95,101

Greenbrier Coal Seam, ... 20

Hackney, Tom, ... 55

Huffman, Clarence, ... 5

Humphrey, Hubert H., ... 82

Imperial Smokeless Coal Company, ... 52

129

Island Creek Coal Company, ... 99
Jeffery L-100 (mining machine), ... 29
Johnson, Lyndon, ... 82
Kennedy, John F., ... 17
Kennedy, Robert F., ... 82
Keyser (mine car), ... 21
Kincaid, Alfred, ... 100
King, Dr. Martin Luther, Jr., ... 17
Leivasy, (W.Va.) Boy Scouts, ... 99
Lilly, Jennings, ... 37,38,70,84,85,88,98,99
Little, James, ... 100
Lynch, Larry, ... 37-40,65,89,94,96-98
mantrip, ...21,26,27,92
Mauls, Jack, Lee, ... 101
Marlinton High School, (W.Va.), ... 16
Marshall University, ... 17
Martin, Eugene H. (Gene), ... 33,34,36,38,66,87,88
90,96,97,98
Mary's Beauty Salon, ... 9
Maust Coal Company, ... 77,99
McCarthy, Eugene, ... 82
McClung, Helen, ... 95

McClung, Renick F., ... 5,37,94

McClure, Paul, ... 79

McClure, Roy L., Jr., ... 43,79

McClure, Roy L., ... 43,79

McKenzie mine, ... 30

McKenzie, Eugene, ... 30

McKenzie, Thomas, ... 30

Meadows, Paul, ... 100

Monica, Sister Mary, ... 79

Monongah #6 mine, ... 26

Monongah #8 mine, ... 26

Monongahela National Forest, ... 11

Moore, Gladys, ... 9,52,85

Moore, John Jr., ...9,10,*13*,19,23,35-37,48,52
58,62,66,72,84,85,91,92,94,96,99,102,103

Moore, Stefanie, ... 9

Moore, Vicki, (Rose), ... 9,10,35,53,55,60,66,68,69
71-73,85,93

Mountaineers, ... 18

Mt. Urim Babtist Church, ... 99

Mullins, Ralph, ... 4,5

National Aeronautics & Space Adm, (NASA), ...17,82

National Forest Service, ... 60

New River Coal Company, ... 27

North Vietnamese, ... 16

O'Dell, Elwood, ... 42,76

pan, (as measurement), ... 32

Perry, Clyde, ... 91

Rader Flying Service, ... 100

Rader, Gerald, ... 100

Red Cross, ... 80

Richardson, C.E. (Charles), ... 83,99

Rivituso, Dinah, ... 56,63

Rivituso, Joe, ... 56

Rudd, Edward F., (Bozo), ... 42,46–50,57,58,70

S & C Coal Company, ... 86,88

Sacred Heart Hospital, ... 79,93

Salvati, T.A., (Tim), ... 47,75,89,100

Saxsewell No. 8 Mine, ... 8,13,18,20.22,24,30,31
*51,*52,56,61,72,101

Scarboro, Edward F., ... 37,38,40,66,84,86,90,96,97

Seabolt, Hershel E., ... 43

Sewell Coal Seam, ... 15,30

shaft mine, ... 22

Short, Denver, ... 100
Siltix Mine, ... 27
Simmons, Roy, ... 92,101
slope mine, ... 22
Snyder, Lowell, ... 101
Southern Mine, ... 75
Stanley, Kermit, ... 91
Straley map, ... 61
Sugar Grove Coal Company, ... 30,98
Sundstrom, H.E., ... 68,77
swag, ... 70
TET Offensive (Vietnam War), ... 16
Todd, Dr. Lee B., ... 75,78
Travis Air Force Base. 80
Tridelphia Mine, ... 26
United Mine Workers, ... 79
United States Bureau of Mines, ... 90,91
Upper Buckeye Presbyterian Church, ... 85
Valley Camp Coal #3 Mine, ... 27
Vietnam War, ... 16,17
Walkup, Eli, (Edward), ... 5,29,*37*,95

Walkup, Hilda, ... 5,95

Walkup, Shelly, Adkins 95

Walton, Ottie J., (Junior), ... 42,47,49,77

Walton, Andy H., ... 42

West Virginia (Vietnam War casualty rate), ... 16

West Virginia Department of Mines, ... 96

West Virginia State Highway 20, ... 11

West Virginia University, (WVU), ... 17,55,56

West Virginians, ... 12,16

Western Union Telegram, ... *95*

Williams, John Ray, (Chipper), ... 16

Workman, Elmer C., ... 52,61

WVAR Radio, ... 55

Made in the USA
Monee, IL
30 October 2020

46390268R00080